Battle Carried

Imperial Japanese
Tiger Art Good Luck Flags
of
World War Two

ISBN: 978-1-943492-57-2

Book design by designpanache

ELM GROVE PUBLISHING
San Antonio, Texas, USA
www.elmgrovepublishing.com

Elm Grove Publishing is a legally registered trade name of Panache Communication Arts, Inc.

Battle Carried

Imperial Japanese Tiger Art Good Luck Flags of World War Two

Michael A. Bortner

EGP

This book is dedicated to my brother, Spec. 4 Thomas W. Bortner, United States Army, 1967-1969. Republic of South Vietnam, 1968-1969. When the burden of technology became frustrating and threatened to temporarily derail this project, you were there to listen to me and offer your support. Your invaluable lift, especially with the illustrations, insured the book's success; I am forever grateful.

This book is also dedicated to Corporal John T. Egger, United States Marine Corps, 1965-1969. Republic of South Vietnam, 1966-1967. No matter where I was in the writing of this manuscript you offered near-daily assurance, suggestions and fine contributions to the content. You defined the word "Friend" and you are missed.

The importance of academic mentors cannot be understated. While it has been a number of years since graduate school, Dr. George Miller provided a good deal of the foundation that I needed to write this book. George, you played an important role in helping me to understand that the material culture, "the stuff", has a voice. We only need to listen and it will speak to us.

Finally, this book is dedicated to Dr. Gerald White. Your undergraduate history lectures were content heavy and old-school. No matter the subject, I enjoyed the minutia, and you never seemed to mind when I recounted it often during your office hours! Thank you for showing me that I could write with passion and still be true to the subject.

John 3:16-21 and Philippians 4:13

Contents

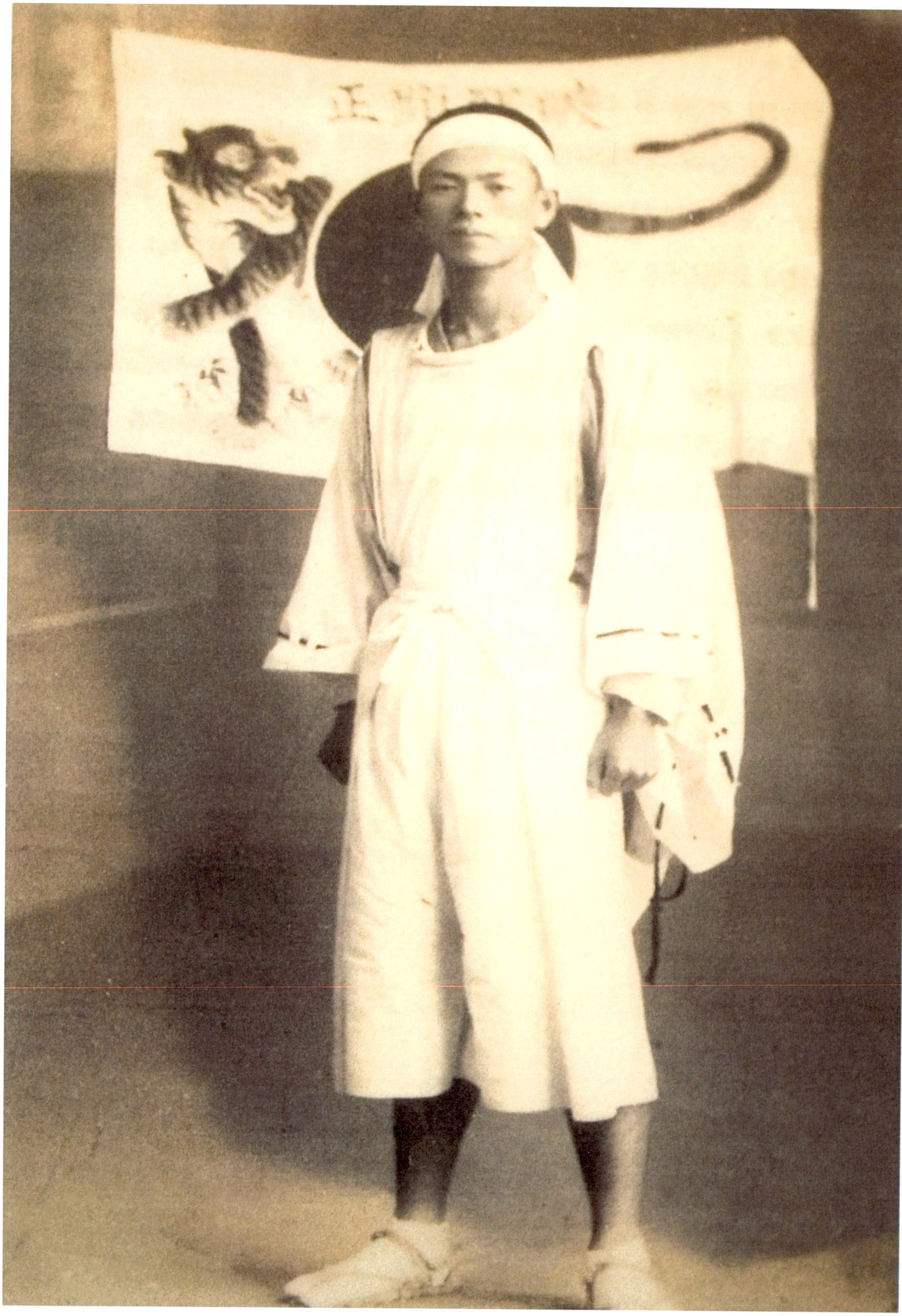

"Looking to thine own duty thou shouldst not tremble; for there is nothing more welcome to a warrior than righteous war offered unsought as an open door to heaven."

– *The Bhagavad Gita*

Foreword

by Carla L. Funk
Executive Director and Chief Curator of University Museums,
Florida Institute of Technology

I first met Dr. Michael Bortner in Melbourne, Florida shortly after the Ruth Funk Center for Textile Arts opened on the campus of Florida Institute of Technology in 2009. The 5,450 square foot Center preserves and displays an international collection of textiles that highlights items from the 16th through the 20th centuries; its primary focus is on items from: Japan, India, Central Asia, Africa, Europe and North America. The Ruth Funk Center for Textile Arts is the only center of its kind in the state of Florida and one of very few in the nation. In Dr. Bortner, I encountered a passionate collector and researcher, who was excited to share his personal collection of Imperial Japanese military textiles with the larger Melbourne audience, the academic community and supporters affiliated with the Ruth Funk Center for Textile Arts.

The Bortner Collection of signed Japanese flags *(hinomaru yosegaki)* and "one-thousand stitch belts" *(senninbari)* was immediately fascinating to me. These dramatic, symbolically charged objects that accompanied men off to war were believed to have protective or apotropaic powers; more significantly than being good luck charms carried into battle, the textiles often celebrated military triumphs or served as reminders of the tragedy of war. As an art historian long interested in the history of material culture- items created for everyday use or ritualistic purposes- not only was I impressed with these items as objets d'art, but I also recognized these textiles as powerful communicators of social meaning.

Equally impressive as the textiles was Dr. Bortner's scholarly devotion to his collection. When we met, he had recently published his 2008 book, *Imperial Japanese Good Luck Flags and One-Thousand Stitch Belts* which provided illuminating details about the use, significance and meaning of military textiles, a subject that, like much material culture, had been neglected by historians. This book became an invaluable source of information for the exhibition that the Funk Center staff and I developed from the Bortner Collection and mounted in the fall of 2012: *Battle Worn: Masculine Ideals and Military Identity in Modern Japanese Textiles.*

Using Michael's scholarship with regard to Japanese symbolism and iconography as a starting point, the *Battle Worn* exhibit examined Japanese good luck flags, off-to-war celebration banners, "one-thousand stitch belts", uniforms and toys from the Bortner Collection and revealed how national duty and familial honor played an essential role in shaping the perception of Japanese soldiers in the 19th and early to mid-20th centuries. The exhibit aligned seamlessly with the Funk Center's mission to promote the cultural awareness and appreciation of textiles. The Funk Center is indebted to Michael for introducing the rich world of Imperial Japanese imagery and its use in military textiles to our dedicated museum members, as well as the curious audience at large.

In this, his second book, *Battle Carried: Imperial Japanese Tiger Art Good Luck Flags of World War Two,* Dr. Michael Bortner continues his iconographic studies with a valuable investigation of tiger imagery, almost exclusively within the context of Japanese signed good luck flags. The tiger is a popular subject in Asian art but it is particularly striking when encountered in Imperial Japanese textiles and militaria. This book provides a survey of the symbolic meaning and cultural backdrop of the tiger from its ancient Chinese roots as a divine creature, to its contemporary appeal as a talisman of good health and protection. *Battle*

Carried examines how tiger iconography developed, evolved and inspired the Japanese public during peacetime and war.

Using rich sources from popular culture that include: Edo era woodblock prints *(Ukiyo-e)*, popular folklore, Zen poetry and Samurai texts, Michael brings the image of the tiger to life and displays the centrality of it, primarily within the framework of Japanese military ideals of victory and strength. I would expect this volume to find a broad readership, particularly among those with a fascination in Japanese cultural anthropology, art history and textile related fields. For the military historian and collector, this research provides yet one more valuable layer to the scholarship that emerged following the release of Dr. Bortner's first work.

Carla L. Funk
October 9, 2018

Preface

One of the greatest pleasures that I have had in over fifty years of collecting Japanese signed flags and one-thousand stitch belts, was when I had the opportunity to write a book on those two subjects. The release of *Imperial Japanese Good Luck Flags and One-Thousand Stitch Belts* in 2008 brought detailed investigation; organization; and photographic recognition to two popular yet under-researched aspects of vintage Japanese World War Two era material culture. The 286-page volume was embraced by collectors and historians of Japanese militaria who, over the years, had perhaps exhibited a *hinomaru yosegaki*, or *senninbari* among their collections, but whose appreciation of this Japanese military send-off-gift was likely limited by the available access of knowledge on the subject. At that time, information on good luck signed flags and one-thousand stitch belts was incomplete at best. Some of the details being circulated were hearsay and often inaccurate. A book that provided research and documentation of the basic history behind this celebrated wartime practice was needed.

From a personal standpoint, the study of flags is a subject that I pursued from a young age. With respect to Japanese signed flags, I remember seeing my first example while engaged in a Scout project. The appreciation that I came to have for these historic wartime relics in time extended to one-thousand stitch belts and other forms of *senninbari*, especially those decorated with stitched, or painted tigers. The enjoyment that I received from the examination and study of these two subjects has evidently rubbed off on others. I am gratified to have received compliments from readers around the world who tell me that their own interest in these topics grew with each turn of the page of that first book. As an author, to receive even modest recognition from that preliminary endeavor makes the time and effort in compiling this new volume all the more worthwhile.

Returning to the subject of the artifacts themselves, we now know that nearly every Japanese soldier, sailor, or airman carried one, both, or multiple examples of the aforementioned items with him as he exchanged the shelter of home for the uncertainties of war. Celebratory, bellicose, humorous and sometimes even thinly veiled risque messages, inked upon a white field by family, friends, or co-workers are what we normally find written upon the front of a good luck flag. Delicately patterned knots, believed to hold within them the assurance of good military fortune and physical protection are likewise visible upon the cloth of a one-thousand stitch belt, or other *senninbari* article. As he set off for war, these personal parting gifts were meant to remind each man of his loved ones back home, and of his duty to uphold the name of his family and the honor of the Emperor.

During the early stages of the Second Sino-Japanese War (1937-1945), Japanese men either volunteered or were inducted into the military and served for a period of two years, subject to recall from the reserves up to the age of 40. In the early years of the conflict and following conscription customs that dated to 1877, once their commitment of service in such places as China, Korea or elsewhere was satisfied, these veterans triumphantly returned home from the battlefields, their military duty nearly complete.

As the war expanded into the Pacific following the start of hostilities with the United States, the need for manpower became critical and then desperate. In the fall of 1943, military service laws were amended in order to access and retain a greater supply of men. At war's end, these regulations were further changed in order to accommodate the use of women for homeland defense. In the minds of the Japanese populace, only two outcomes to the conflict seemed likely: either complete victory or the unthinkable, total defeat.

As part of the warrior's code, or *bushido*, surrender was unacceptable to the people of Japan. Death on the field of battle demonstrated that the soldier had given all that he had in the performance of his duty. This sacrifice brought great honor to both family and one's ancestors. As the fortunes of war grew critical, the logical outgrowth of this ideology was the reality that no man was expected to return home from overseas service. Following the ancient custom of the samurai preceding death, a body was not necessary for a funeral. Japan's modern fighting men of the 20th century normally left behind specially designated envelopes that contained lockets of the man's hair and/or fingernails that were intended to be used by the relatives should a funeral service be necessary.

The disastrous end of the war left those remaining former Japanese service personnel and the civilian population back home with a stark reality: their nation was in ruins. The result of their ambitious attack upon the United States of America at Pearl Harbor, championed by the hard-line militarists and nationalists, left post-war Japanese citizens with few choices. They were led to re-build the economy, create a defensive military only and put any memory of the war, especially those items of a glorious wartime material culture, into either the storage shed or the trash.

Across the waves to the United States, the war's end in 1945 brought forth a completely different reality, evolving post-war culture and history. Nine months after the end of hostilities, "[The] cry of the baby....[was heard] across the land." This trend, which author Landon Jones described in his book, *Great Expectations: America and the Baby Boom Generation* would not end until 1964. Within that short time frame, nearly 76.5 million children were born to American couples both young and old. Eventually those surging statistics would grow to constitute nearly 40 percent of the United States' population. It was into this exuberant period of possibilities that many of my friends and I were born.

In the early 1960s, virtually every family had a father, uncle, or grandfather who served in the United States military during World War Two (1941-1945). After all, more than 16 million American men, 10 percent of the population, had picked up arms in defense of their country following President Roosevelt's declaration of war, subsequent to the Japanese attack on American military and civilian populations at Pearl Harbor. In those post-war days, boys my age compared notes regarding relative's military service. In fact, there came to be a pecking order of sorts among fellows, from which bragging rights were derived based upon a number of factors: father (or relative's) branch of service, his rank and/or job in the military and the location in which he served or saw action. Growing up along the California coast, it seemed to me that more of my neighbors were veterans of the Pacific War than those who served in the European Theater. If the stories circulated by friends on the ball fields of my youth were any indication, then this perception was reality.

Many of those amazing first-hand exploits became even more personal later when, after a busy day, a friend might invite a group of us over for dinner so that together we could plead with their fathers to recount for us their not so long ago wartime adventures. As spine tingling as some of those recollections seemed, I am positive that the actual realities had already been heavily tempered; a dark stare from a buddy's mother toward her husband across the table, being enough to bring even those said stories to a firm halt.

Nevertheless, the verbal recounting of combat memories gave us personal contact with the actual witnesses to those historic events. Places like Tarawa, Saipan, Guadalcanal and Iwo Jima were all familiar to us. Within the context of the "Baby Boom" generation especially, this storytelling had the effect of imparting values that many of us still embrace today: hard work, self-reliance, perseverance and individual effort that contributes to the team's success.

I recall an instance when one of those dramas ended with a surprise that had a huge impact on me. One evening after dinner, a group of us were invited to sleep over in a neighbor's backyard. This family owned a tent, purchased from the local Army-Navy store. Using it to set up our bivouac, my group of friends pitched the leaky tarp in the early summer, and there it remained until just before school began that next fall. This dilapidated canvas acted not only as our clubhouse, but served as a great place for our cadre of friends to get together and hash out all of the important topics of the day.

This particular night, some of the guys assembled in order to compare notes on a movie re-run of the John Wayne film, "The Fighting Seabees" seen recently on the popular "Million Dollar Movie" television show. As early evening began to wane and attention spans followed suit prior to lights out, a voice was heard from out of the falling darkness. Our host's dad hailed the tent from across the far side of the lawn, requesting "permission to come aboard".

During the war, my friend's dad served with a Marine regiment as part of the early diversionary force at the Battle of Saipan, in 1944. As he recounted that event, he told us that he felt fortunate to have been a part of the later wave of troops that landed on the beach just behind the initial attack on that first day. He recalled how tough the battle was on the entire group of Marines because of the accuracy of the Japanese gunners, who poured pre-measured fire down upon the beaches from their superior positions.

Next, he recounted for us one of his most frightening memories: The Japanese counter attacks that took place during the first evening ashore. We listened breathlessly to the story. Later, he asked if we would like to look over some of his souvenirs from the Pacific War. A number of the items had been picked up on Saipan that second day, as he and a few of his fellow Marines walked sections of the beach in front of their fox holes and surveyed the ghastly site before them. He told us that his wife did not object to his keeping a few things in his old sea bag, as long as they remained in the attic of the family's garage.

Having been buckled up for at least 20 years, the duffel had acquired the scent of salt and mildew, which escaped when opened. The first item that he removed from storage was his Marine Corps uniform, followed by a stack of letters that he mailed home from his training days and time overseas. An old family photo album was the next item to emerge, followed by some miscellaneous other souvenirs and certificates from shipboard cruise ceremonies to various far-away places. Finally, some real Pacific War "bring backs" made their way out of the bag and into the beams of our unsteady flashlights.

I remember seeing what I later learned was an *Arisaka* bayonet and scabbard. Next up was a Type 90 steel helmet, still wrapped in its protective cloth clover. He shared with us other items including a few sake cups, bank notes and some old black and white pictures of Japanese men in service uniforms. The next item that really caught my eye, however, was an impressive battle flag. Emerging from the bag after all of those years, it seemed to resist the effort to unravel it. Nevertheless, after some coaxing, my friend's dad was able to shake open the folded piece of cloth. Numerous *kanji* characters inked along rows of red and white rays ran up, down and around the flag's length. At the time, the lines reminded me of soldiers arranged in neat formation, ready for inspection.

Of course, recollections like this create as many impressions as there are people. Along with the memories of acquiring my own first good luck flag and one-thousand stitch belt, the signed battle flag that I examined that night and the story of its recovery were branded upon my subconscious. The reality was that in those days, exchanges like this one were fairly typical of the time. While scores of veterans remained stoically silent about what they saw and did during the war, others took advantage of the interest among their young audiences and used the dramatic moment as a cathartic release that enabled them to cope with the frightening memories. This had a huge impact on us kids too: although once removed from the actual events, we shared the history albeit in safety; re-lived through the eyes, ears and mouths of our relatives and friends.

It is my desire that *Battle Carried* catalyzes a broad study and invites discussion in Japan, the United States and elsewhere with regard to those momentous events that gripped people worldwide more than seventy years ago. While most of those men and women who gave of their youth and lives have passed into history, quite often the material culture that they carried forth or recovered, remains. Those relics of the war have the potential to speak honestly to us of their struggles, deeds and valorous actions; for this I am thankful.

On a similar note, I must extend my gratitude to the private collectors who care for some of these relics; pieces earnestly sought and gathered from faraway places, or passed down from relatives who lived the history. At times disparaged, their role is vitally important: they are often the only ones willing to spend the requisite resources in order to acquire, preserve, study and share with others the history therein. We owe them a debt of gratitude as they seek to secure these artifacts- items often found gathering dust, unremembered in some museum back storage room, or worse, relegated to a dishonorable toss onto the trash bins of time.

Michael A. Bortner
Belvedere, CA
May, 2018

Introduction

The purpose of this book is to expand upon a subject that garnered enormous interest in my first volume on *hinomaru yosegaki – Imperial Japanese Good Luck Flags and One-Thousand Stitch Belts* – good luck art flags. Such flags, particularly those that depict defiant, snarling tigers in varying poses, continue to be some of the most popular objects discussed among collectors of Japanese World War Two era signed flags and militaria. While much of the content of the characters upon the fabric itself is usually of the more typical good luck variety, covered with the signatures of friends and accompanied by patriotic slogans, the various renditions of tigers upon the "cloth as canvas" are quite often superb and make those rare examples anything but ordinary.

One of my aims, therefore, is to introduce the reader to a deeper understanding behind some of the aspects of Asian tiger iconography: not only how it developed, evolved and motivated those who served in the Japanese military, but its general influence as an artistic subject across Asia as well. Whether the reader is an expert in the subject of military related flags matters little. Indeed, the quote by the well-known abstract painter Darby Bannard applies here: "You don't have to be a caveman to appreciate Lascaux."

Of course, knowing that there is more to almost any pursuit than what meets the proverbial eye should bring a broader acknowledgment to nearly anyone who chooses to read about this subject. It is that spirit of appreciation that Jacob Bronowski, author of *Science and Human Values,* referred to when he said, "[We] live again the moment when the creator saw and held the hidden likeness." In the context of World War Two era signed flags and their Shinto and Zen Buddhist influences, Bronowski's quote conveys the sentiment that the artistic depictions of tigers, even absent their many anthropological circumstances, are stunning for the manner in which they capture animistic and/or anatomical reality.

Additionally, Japanese textile art aficionados and students of *nishiki-e* art alike will enjoy drawing comparisons between some of the influential styles of the legendary *Ukiyo-e* masters from the distant past and the typically anonymous mid-twentieth century tiger flag artists listed among these pages. After all, highlighting with images these evolutionary artistic correlations are a worthwhile endeavor of the book; a study the likes of which has arguably never been done. It is the author's hope that a thoroughly detailed analysis of these relationships emerge in the future.

Whether executed by artists in muted shades of black, gray and white or in combinations of brightly mixed colors, the Imperial Japanese tiger art good luck signed flag will remain one of the most popular varieties; a quintessential symbol of the Japanese fighting spirit of the era. Symbolically woven into the fabric of Asian culture and appreciated by others as well, tiger imagery will continue to capture the imaginations of viewers everywhere.

1. Tigers And Japan's Ancient People

"A man's good name is as precious to him as its skin is to a tiger." – *Japanese Proverb*

Clear fossil records that date from approximately 1.6-1.8 million years ago, indicate that the big cat known as the Javan tiger *(Panthera tigris sondaica)*, would be the oldest, recognizable tiger species today if it were still alive. Fossils of more ancient tiger species *(Panthera zdanskyi)*, have been uncovered in China and date to more than two million years ago. By the end of the Pleistocene epoch (2.6 million-11,700 years ago), the tiger's habitat extended across portions of northern Asia, India, sections of Beringia and Russia's Sakhalin Island. The Islands of Borneo in Malaysia; Sumatra in Indonesia; Palawan in the Philippines; Java; as well as others, came to be inhabited throughout that era by their own tiger species. While there is a lack of consensus that all of the previously mentioned islands were once inhabited by tigers, legends and folk lore as well as extant dental artifacts from some of those places have proven hard, over the years, to ignore.

Modern computer models have shown that the Islands of Japan should have made ideal tiger habitat. No matter "that evidence", as it seems that there is little in the way of artifacts to support that tigers ever lived there. Keeping the "Japanese Tiger" debate alive within the last few years was the discovery there of a tiger-related species, or perhaps two: *Panthera tigris palaeosinensis*, and *Panthera youngi*. Genetic studies on the fossils clarified whether these cats were actually tiger related. In fact, the first animal was more closely related to the lion and/or leopard, while the second, re-classified as Felis youngi, showed a hybridization of both lion, and tiger characteristics. Neither, strictly speaking, could claim a clear connection to our modern species of tiger. As the years pass and the tiger's genetic waters prove even less murky, there are growing numbers of experts that agree that it is unlikely that the Japanese Islands were ever inhabited by full size tigers *(Panthera tigris tigris)*, similar to those animals currently observed on the Asian mainland, and elsewhere.

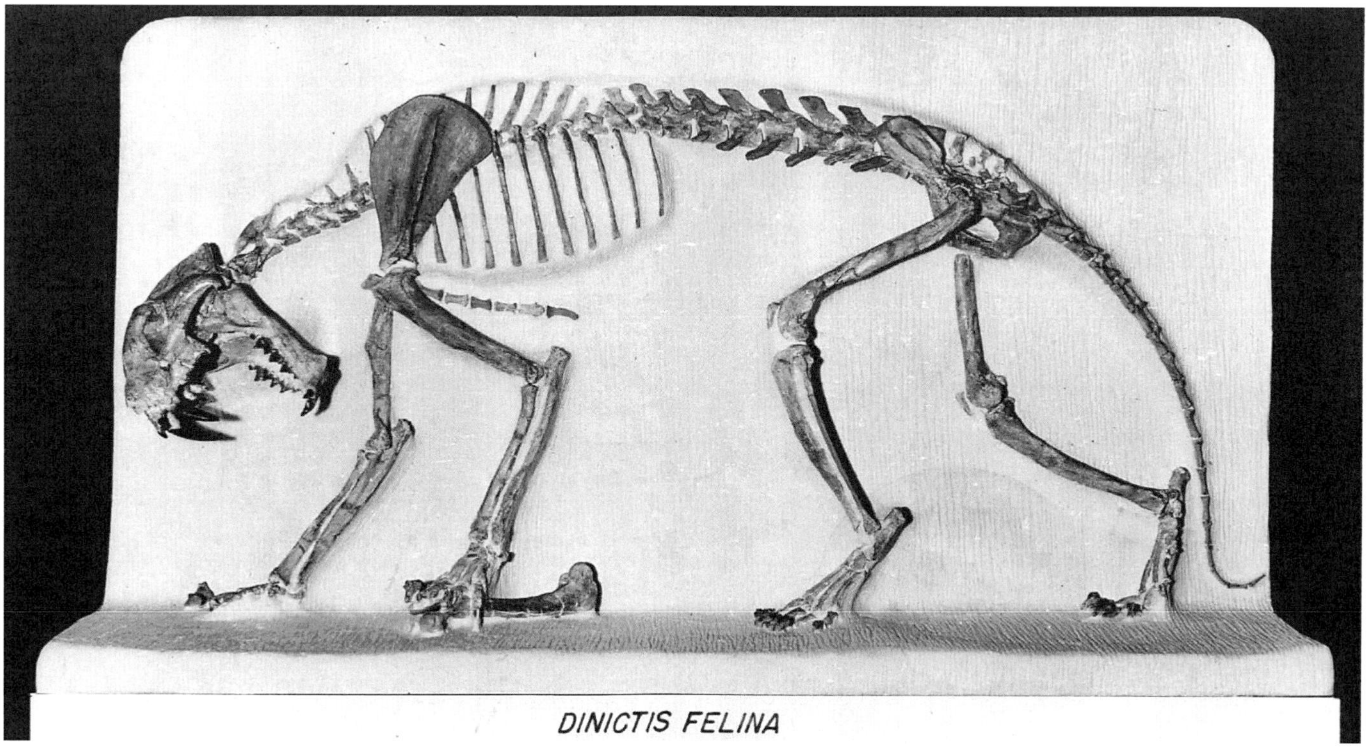

Familiar fossil remains of the striped tiger's wholly unrelated "tiger-cousin," the Saber-toothed cat *(Smilodon populator)*.

The other varieties of tiger include: the Indian, or Bengal tiger; South China tiger; Malayan tiger; Indo-Chinese tiger; Bali tiger (extinct since the 1940s-1950s); Javan tiger (extinct since the 1970s); Caspian tiger (extinct since 1970).

During the Pleistocene (2.6 million-11,700 years ago) and Holocene (11,700 years ago-present) Eras, Japan's three largest islands: Honshu, Shikoku and Kyushu formed one huge land mass identified as Hondo. During these early periods of maximum glacial formation, Hondo was attached to the East Asian continent. Later, during the last glacial period, no such connection was present. Rather, ice bridges existed between Hondo and Hokkaido Island, and these allowed for the migration of many animal species, as well as the human hunters that followed them. Coincidentally, Taiwan to the south of Japan was attached to portions of the Asian mainland, and extended out to connect with the Yaeyama, and Miyako Groups of Islands, as well as others in Japan's Ryuku Island chain.

Today, *Panthera tigris* has nine sub-species; three of which are extinct. Of the six remaining species, the Siberian tiger (shown above), is the largest, with males weighing as much as 660 lbs.

The smallest tiger species, the Sumatran tiger. Males weigh in at approximately 310 lbs.

Glacier cover in Japan (Hondo and Hokkaido Island), at the height of the last glaciation about 20,000 years ago. *(See appendix 1, page 192).*

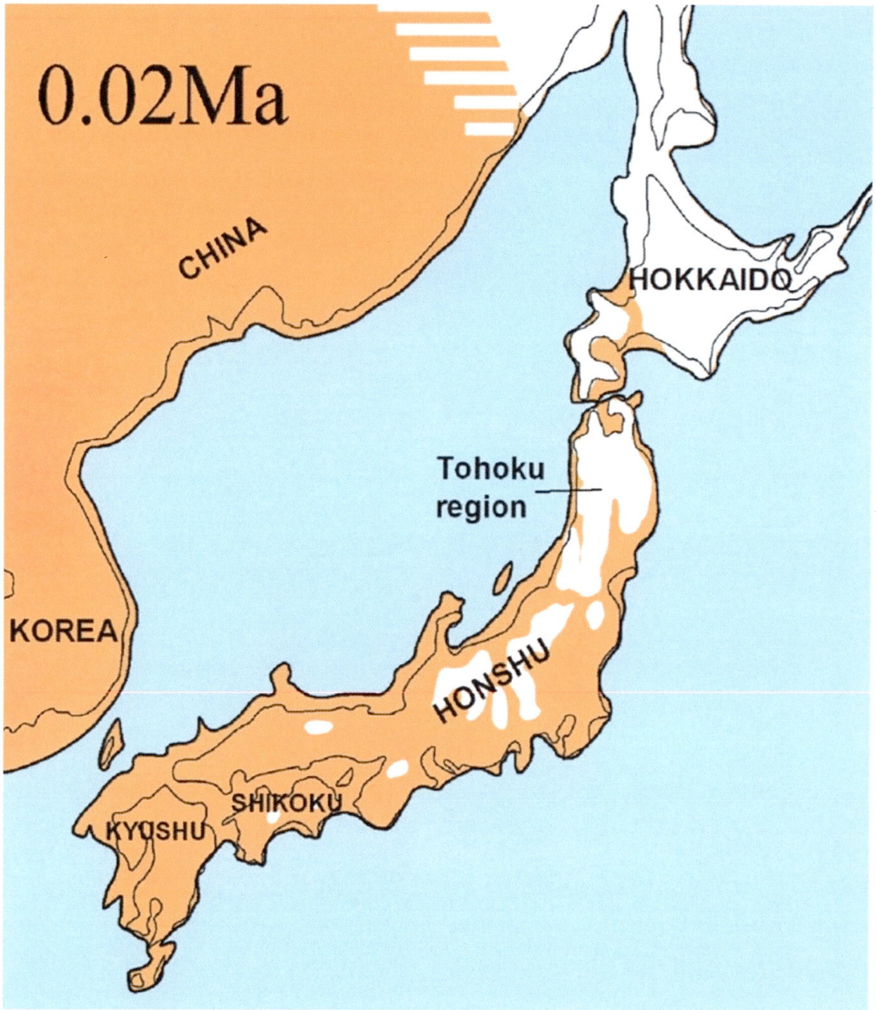

On Hondo there is some evidence to suggest that humans had reached the Japanese Islands approximately 30,000-40,000 years ago. Many evolutionary biologists, and geneticists now generally agree that the Japanese people descended or were blended from native Jomon, and non-native Yayoi populations that lived there in proximity to one another.

A key question has always been, "How did early humans find themselves in the land that would eventually become Japan?" Most scholars today believe that Japan was inhabited in two distinct waves: the first wave occurred approximately 40,000-12,000 years ago, while the second wave may have taken place as recently as 2,300 years ago. It is highly likely that early man boarded small seagoing boats and traveled there via the linked chain of islands between Taiwan, Okinawa and Kyushu, Japan's southern most island. Debate as to additional migration routes of access to Hondo/Japan are ongoing:

In General:

1. Indigenous native groups gained access to available land as the the glaciers receded; Southern native populations extended themselves north from the Ryukyu Islands, while northern groups, possibly from Russia, pushed southward into Hokkaido.

More Specifically:

2. Migration may have allowed populations from Korea, and Mainland China to reach Kyushu along a southern route. To the north in Russia, entry may have been made from Sakhalin Island to Hokkaido, creating travel along a northern route.

3. Additional migration of people groups from the Pacific Islands, portions of Southeast Asia and possibly India may have occurred as well.

As the Pleistocene Era came to a close, the ice flows receded, and sea levels began to rise. Well used land bridges, and similar travel routes that were accessed in order to pursue wild game were inundated. The remaining inhabitants found themselves

Japan today in relation to Korea, China, Russia, and its southern island neighbors. *(See appendix 2, page 192).*

island bound, and over time came to form the genetic core of today's Japanese people.

As time passed, the earliest aboriginal people faced few choices: they either died out, assimilated voluntarily into the dominant, early-evolving "proto-Japanese" culture, or were forced into it. One of the largest of these last aboriginal groups (25,000-200,00 people), the Ainu, continue to this day to live primarily in the Hokkaido region of northern Japan.

Ainu men share a meal (c.1910-1920).

2. The Chinese Painting Style and its Influences Upon Japanese Art

"The supreme art of war is to subdue the enemy without fighting." – *Sun Tsu*

As one of the world's oldest continuously used artistic traditions, the Chinese Painting style has had a heavy influence on cross-Asian art, for an extraordinarily long time. While exact dates for the lengthy period are discussed, a fairly accurate time-frame would begin with the *Han* dynasty in 206 B.C., and extend through into the Late Imperial China period, ending in the *Qing* Dynasty around 1895 A.D. Because of its important influences upon Japanese art, a brief understanding of its development and effects is necessary.

The earliest forms of Chinese painting depicted animals and designs that were used to decorate pottery, lacquerware and other utilitarian items. As the style slowly evolved, later generations of artists built upon successive techniques pioneered by their artistic forebears. As their skills developed and matured, Chinese painters began to represent more sophisticated subjects. Among other themes, landscape art that incorporated mountains, water and plants became extremely popular. Of particular note: animals, and human figures, in relation to their outside world were added into these often idyllic portrayals.

The basics of Chinese painting incorporated long practiced and deeply admired calligraphy techniques that relied upon brushes dipped into water, followed by the application of black ink and/or colored tints; the use of oils was omitted. For years, this art form was utilized to create beautiful silk wall or floor screens as well as other decorative objects. The use of silk as "canvas" proved quite expensive. Eventually, a reliable, inexpensive medium had to be sourced in order to replace it.

The chroniclers of the 5th century *Han* dynasty (206 B.C.-220 A.D.) took note that it was the Chinese court eunuch, *Ts'ai Lun* who invented, and made later improvements upon paper in the year 105 A.D., using mulberry, rags, tree bark, hemp and other materials. Much older paper fragments would later be discovered by archaeologists in a Chinese tomb, revealing that paper had likely been in development quite earlier than previously recognized.

Following its use in China, paper eventually made its way to Korea in the 4th century. Afterward, handmade paper or *tesuki washi* seems to have been introduced across the sea to Japan by the Korean Buddhist monk, *Doncho* around 610 A.D., although some sources indicate possibly as early as 280 A.D.

Two fundamental approaches to Chinese painting would eventually develop: The first is known as *Gongbi* and means "precise" or "exact". This highly detailed, colorful form relies upon the fastidious execution of brush strokes. The subjects of *Gongbi* normally take shape as narratives or stories, and frequently involve figurative forms. Artists that mastered this approach were often skilled craftsmen, employed within the Courts and/or who operated independently due to the demand for their products.

The second approach is known as *shui-mo*, which translates to "ink and wash" or "brush and water". Sometimes called "literati painting", the form was practiced by those with the requisite time and sophistication necessary to master its rigorous system. Chinese "ink and wash" found express favor among academics and cultured gentlemen who highly valued it as one of their four key artistic approaches, the other three being: the *Guqin*, a seven-stringed instrument known as the "zither of the sages"; the two-person board game known as *Go*, whose objective is to surround more territory than your opponent; and Chinese calligraphy.

With the practice of Zen Buddhism taking hold all across East Asia, this aspect of Chinese painting took on primary significance. The key to understanding this approach was that the artist must not only be able to depict his subject in a life-like manner, but more importantly must endeavor to capture its essence, or spirit, within the painting. This was accomplished in a free-hand manner through the application of different thicknesses of ink, light and heavy brush strokes and masterful shading, all of which the student/practitioner must first practice, over and over, to rote perfection. In each case, brush strokes were viewed as the crucial manifestation of the artist's inner spirit.

The first historical encounters that the Japanese people had with the tiger came in the form of art which dates back to the 7th century and perhaps even earlier. Cultural interface with China and Korea, where the tiger was plentiful, are the most likely source of this exchange. Japan's earliest cache of Chinese tiger related paintings, and artwork were first recorded in the collection of the *Ashikaga Shogun* in the latter part of the 15th century. Also known as the the Muromachi shogunate because his palace was located in the *Muromachi* District of Kyoto, that ruler-ship was established in 1336 and lasted until 1573.

Over the years, the *Ashikaga* ruler traded most heavily with his international neighbors, primarily the Mongols, and later the *Ming* Chinese, as well as the Koreans. Key influences of those exchanges, the Zen Buddhist concepts of nature, beauty, and simplicity, spread through Japan, particularly among the samurai leadership. Those ideals had far reaching affects upon various aspects of military strategy and philosophy, architecture, design, poetry and art.

Early Japanese paintings, and carvings fashioned by Zen Buddhist priests upon decorative screens, scrolls, wood doors, and panels were only some of the aspects of religious material culture where tigers were highlighted.

This Edo period woodblock print by the artist Kikukawa Eizan (1787-1867), is titled *Artist Painting a Tiger on a Screen.* Eizan first studied under his father, the *Kano* School painter Kikugawa Eiji. Influences of the skillful master Utamaro (1753-1806), may also be seen throughout his work.

By the 16th century, Japanese artists began to create their own tiger images by copying works of art imported from the Chinese and Koreans. Having never actually seen real tigers made this a stretch in the beginning for the Japanese, who also used the imported pelts of the animal as a visual guide for their paintings and drawings. Making the task easier, but producing an even less accurate representation, occurred as some artists simply copied the tiger's orange and black striped pattern onto the painted images of the familiar house cat. One more faux pas of this early period involved the leopard, which was also indigenous to the Asian Mainland, and pairing it with images of the tiger under the assumption that the smaller leopard was simply a female tiger.

Perceptions, and realities more closely aligned as the 16th century faded and the 17th century dawned. Later (some sources say the mid-1800s while others give much earlier dates), actual tigers were introduced to Japan, having been captured live overseas and put on display in the royal court and elsewhere.

Some of the most influential tiger artists of the 17th century came from the Japanese Kano School, where long developed creative patterns of thought, coupled with the popular Chinese style of painting, acted upon those artists and their artwork. As one of the predominant styles of its time, lasting from about the mid-1400s until approximately 1868, the Kano School and its atelier helped produce art that the public came to admire and add to their homes and private collections in large quantities.

The style took its name from Kano Masanobu (1434-1530), who as master, established his professional studio in Kyoto. Kano's family members, students, and associates created works of art that were delivered to patrons, and customers under his name. While Masanobu was not a Zen devotee, he recognized the influence, and financial wherewithal held by the monks within the surrounding Buddhist temples and associated himself with them. Because they preferred Chinese Painting stylistically, Masanobu and others learned that technique to great benefit, and influence.

The production of Japanese woodblock prints, or *Ukiyo-e,* became one of the most dominant forms of art in Edo Japan (1603-1868). Fabrication involved a variety of articulated steps and included a cadre of specialists who provided a specific skill for every aspect of the process. The artist, publisher, woodblock carvers, and printers each performed an important function that were combined to give us the beautiful prints that remain to this day. The popularity of the form, then and now, speaks for itself as many of the images were originally intended to be appreciated and discarded. While a few woodblock "runs" were designated for limited edition distribution, most were mass-produced works of art that were targeted for a circulation of roughly a few hundred copies. Those would then be eclipsed by the next fashionable artist's release, or the next specialty theme. Where demand for a picture was anticipated to be fairly high, "runs" might extend into the thousands of copies, with some of the most popular designs reaching close to twenty thousand copies. If the printing blocks became saturated with ink, they would be set aside and allowed to dry out in order to preserve the former clearly defined inked image. If demand and use was too great, it was possible for the blocks to become damaged. In those instances, the woodcarvers would be re-engaged and the blocks carved one more time, to match as closely as possible the original block's design.

As wealth, and the display of it grew outside of the restricted upper classes of Japanese, particularly during the late Edo period, a growing ability, and demand to show off that affluence found expression in the purchase of art. As interest piqued, there was a similar growth in the popularity of various artistic subjects. Some of those topics included: samurai warriors and famous battles; kabuki actors; beautiful geisha and courtesans; land and seascapes; cosmology; as well as *shunga* or erotic art.

In addition, as further sub-sets of that expanding art movement: themes that portrayed famous samurai as tiger hunters (known as *musha-e*), tiger folklore and legends, tiger imagery printed or embroidered upon the clothing of art subjects, tiger skins or tiger body parts worn as trophies or emblems of status, were all detailed visually by the artists who designed for woodblock print art.

Ukiyo-e print shows the legendary *samurai* warrior, Kato Kiyomasa defeating a tiger. Kiyomasa wears the distinctive *nammyouhorengekyo kamon* ("O"), on his armor as well as on his banner.

Mirror of Magical Heroes is the title of this 1877 woodblock print by Toyohara Kunichika. Each man portrays a famous magician, with the performer in the center wearing a tiger on his kimono. That character, known as *Tora Omaru,* is being played by the famous kabuki actor, Ichikawa Sadanji.

In this hand colored woodblock print of the samurai warrior, Minamoto no Yoshitsune (1159-1189), the handle of his sword *(tsuka),* and saddle blanket may be seen covered in tiger stripe fur.

A sampling of some of the most popular *Ukiyo-e* artists that span the era, and who popularized the tiger's image include: Katsushika Hokusai, Keisai Eisen, Kishi Ganku, Maruyama Okyo, Utagawa Kuniyoshi, Utagawa Kunisada, Tsukioka Yoshitoshi, and Utagawa Yoshitsuya among others.

In this *Ukiyo-e* print by artist Utagawa Yoshikazu (c.1850-1870), the much admired samurai general Yoshitsune is depicted sitting upon a tiger skin rug, indicative of his rank.

3. Tiger Zen

"*Budo* is not a means of felling the opponent by force or by lethal weapons.
Neither is it intended to lead the world to destruction by arms and other illegitimate means.
True *Budo* calls for bringing the inner energy of the universe in order,
protecting the peace of the world and molding, as well as preserving,
everything in nature in its right form." – *Ueshiba, Morihei*

In Asian cosmology, the tiger was seen as a divine creature that played a significant role in how those cultures understood the origin and evolution of the universe. In Taoist art, the tiger was frequently observed representing the "Yin" to the dragon's "Yang". When the tiger *(tora)* was complimented visually with the dragon *(ryu)*, one of the most prolific pairings to illustrate the Japanese Zen Buddhist struggle for enlightenment emerged.

Ukiyo-e print titled *A Great Competition Between Tiger and Dragon in the Wind and Rain* (c.1856), by the artist Utagawa Yoshitsuya (1822-1866).

With some exceptions, the Japanese embraced the zodiac system of the Chinese. The Tao constructs the world around two forces; they operate within a Yin-Yang relationship. Yin characteristics are composed of water, wind, earth and are murky in nature. Furthermore, their essence is female and static. The aspects of Yang incorporate fire, rain, the heavens and brightness. Their essence is male and energetic. The elements described may be manifested in the combined Yin-tiger, and the Yang-dragon; the pairing is known to the Japanese as *uchu no tora*, or "tiger in rain". Zen Buddhism acknowledges an interplay between these two natures, one that exists throughout the entire universe.

The tiger, with its courageous character, is accepted throughout Asia as the most esteemed of all the large wild animals. It holds its third position in the zodiac pantheon of twelve, just behind the rat and the ox. In pictures it is frequently positioned focused, ready to pounce upon its prey. Similarly, it is often portrayed descending along rocky outcropping, its belly stretched out low, hugging the ground. As a common theme, wind-strained bamboo thickets typically occupy the same image as the growling orange and black striped beast.

The late orientalist, Robert van Gulik wrote that, "In Japan, the tiger portrayed among bamboo stalks in the wind is known as *take ni tora*, 'tiger in bamboo'. This representation is generally taken to symbolize that even the most powerful of terrestrial forces, namely the king of all animals, had to yield to the forces of nature. As such, the tiger in the *take ni tora* representation is also said to be identified with the wind itself, symbolizing as it were, the rustling wind in the bamboo grove." The English born barrister, and art collector, Marcus Bourne Huish expounded upon this relationship further when he wrote in his 1889 book, *Japan and Its Art* that the tiger "...is very often depicted in a storm cowering beneath bamboos, signifying the insignificant power of the mightiest of beasts as compared to that of the elements. When merely seen in connection with bamboos, it is so because its power is such that it can transverse a thousand miles at a stride, even through a bamboo forest." The powerful cat has a tempered force that is evident in its rigid muscles, allowing it comfort in its Yin/earth realm.

The dragon typically shows its force in a more spirited manner. He is often rendered, surrounded within the heavens by angry rain clouds and storm energized water or seas. Emanating out of the celestial realm, the dragon is frequently shown descending toward the earth where his Yang menaces but does not dominate, the tiger's Yin. Those two forces, uniformly matched are in balance, as they typify the universe's harmonious nature.

An image from the *Ukiyo-e* print series entitled *Drawings Of Birds And Beasts* (c.1835-1837), by the artist Utagawa Kuniyoshi (1798-1861).

4. Epic Tiger Legends and Battles

"There is certainly nothing more important in life than what we do at the present moment. A person's entire life consists of nothing more than one moment piled on top of another, over and over again. Once enlightened to this, the warrior has nothing else to worry about, because he realizes that he has only to live in the present moment with the utmost intensity."

– *Hagakure: The Book of the Samurai*

The importance of the tiger and the significance of its interaction among human beings within Asian and popular Japanese culture in particular, has been preserved in numerous ways, not only in art but through a combination of oft retold histories, legends and stage plays. In much the same way that early wilderness traditions contribute on the one hand to the personae of the real life frontiersman and fabled bear hunter Davy Crockett, or the larger-than-life American folk legend of lumberjack Paul Bunyan, and his blue ox Babe, were once widely recounted in the United States, a similar number of well-known Japanese warriors have equally exciting dramas that surround them.

Japanese good luck signed flags, or *hinomaru yosegaki* are sometimes graced with an artistic representation of the tiger. The animal's image generally takes various forms, with the most basic, painted upon the flag in black ink. At other times, red from a personal stamp, or *inkan,* expanded the color pallet to three or four shades, often adding hues to the animal's tongue or orange-brown to its coat. In rarer instances, an artist might actually paint the tiger upon the flag in a host of colors and create a full, realistic likeness. Finally, in a few examples, the tiger may have been silk screened, or block pressed onto the front of the flag in much the same way that off-to-war banners, or *shussei nobori,* were created. *Nobori* with tiger images upon them, while of interest, seem to be less popular among collectors, probably because they did not normally accompany their owners overseas, and therefore seem to carry less historical context. Also, the longer *nobori* examples may exceed 20 feet in length and take up a significant amount of space when displayed.

The question naturally remains: How did the tiger come to be associated with the Asian warrior in general, or more specifically, the Japanese fighting man? Numerous cultural icons were passed along to the Japanese after first being introduced from the Asian continent either via India, China or Korea. These influences might initially have been embraced, due to their religious contexts, and may have been modified in order to suit different circumstances or values in Japan. As the years passed, a layering, or evolutionary effect, frequently took place that combined cultural influences from a variety of places. This essentially created an amalgam that finally led to the idea fitting within a wholly Japanese context.

One of the earliest relationships that binds the tiger to the military came about in a story described by the Chinese *Han* dynasty (206 BC-220 AD), historian, Sima Qian. In his epic work titled *Records of the Grand Historian,* he provided details of the Battle of Banquan, which was the first recorded military campaign in Chinese history. Because it was officially noted in Sima Qian's history book, it is thought that this was a factual encounter, marking the actual cutoff point between legend and fact- based history.

Fought between the primitive forces of Huang-di, known as the "Yellow Emperor", and Yan-di, known as the "Flame Emperor", a general picture emerges with regard to the details of the clash. Debate continues to swirl around where the event actually took place, with different historians providing educated guesses. Apparently also lost is the exact date of the confrontation, surmised as having likely occurred sometime in the 26th century B.C.

We know, however, that the victorious forces of the "Yellow Emperor" fought under specific totems; one of those being the tiger. It is believed by some ancient peoples that totems, or tutelary deities, may be actual spirit beings, or guides, emblematic of and to, a certain clan, tribe or people group. Animistic people groups believe that at certain times, they may become indwelled by these spirits who provide them with the specific characteristics of that animal. The Sinologist, Leopold de Saussure (1866-1925), re-enforced this theme when he recalled the story of Huang-di, who was said to have "tamed and trained tigers" to be used in his war against the forces of the *Shennong* tribes, and Yan-di. Based over time upon this cultural interplay, the tiger and the warrior became inseparable. Later, in the second century, fighting men were said to have embraced tigers due to the animal's ability to scare away demons and to hold evil spirits, or the enemy, at bay.

There is a genre of *Ukiyo-e* art known as *musha-e,* that examines important warrior struggles. Within our context, those deal specifically with battles between warriors and tigers. Many names have been connected to well known stories over the course of hundreds of years, and an entire volume could probably be devoted to just this topic. However, a couple of the more popular human-tiger legends focus on the historical samurai warriors known as Watonai and Kato Kiyomasa. Not only do the following stories combine a measure of fact with fancy, but they demonstrate how, and why each of these men traveled to the Asian continent, and fought harrowing battles with the king of beasts. Those encounters proved beyond a doubt the men's cunning and bravery in the face of death!

The Legend of Watonai

The historical figure Zheng Chenggong was a prominent warrior and general, born to mixed Chinese-Japanese ancestry. Known both as Coxinga, or by the more popular name, Watonai, which translates as "between Japan and China", he was born in the Japanese coastal town of Hirado, Nagasaki prefecture in 1624 A.D. His Chinese father, Zheng Zhilong, was a wealthy merchant-pirate who spent his time traveling back, and forth between China and Nagasaki. His Japanese mother was probably Tagawa Matsu, who appears to have raised the boy by herself for his first seven years. The history of that life, and of his family, reads like a long, multi-act play. Indeed, popular theatrical productions were written and performed in Japan to highlight assorted aspects of the family's life. *Ukiyo-e* artists enshrined a number of the play's characters and more famous scenes with their wood block prints.

The most popular of those performances was written as a puppet play, or *bunraku,* in 1715, and titled *The Battles of Coxinga* or "*Kokusen 'ya Gassen.*" Created by the legendary Japanese dramatist, Sugimori Nobumori who became famous under the pen name, Chikamatsu Monzaemon (1653-1725), the play is loosely based upon the historical events of the main character Coxinga/Watonai who, with his mother, travel to China in hope of defeating its rulers and restoring the *Ming* dynasty.

In the illustration below, the acclaimed *kabuki* actor Kawarazaki Gonjuro 1 (1838-1903), strikes an angry pose in his portrayal of the figure Watonai. In a well-known scene, the Japanese warrior expected to enlist the support of General Kanki, who was married to his half-sister Kinshojo, in order to bring about regime change. The colorful woodblock print graphic demonstrates vividly the mouth of the actor, turned down in a scowl upon hearing that the general has refused to assist him. The stage make-up around the eyes and hair of the actor are used in order to express valor and to give him an even more fierce appearance.

As the play develops, Coxinga/Watonai and his mother are seen trekking through China, when they become lost in a bamboo thicket. Soon, a large tiger appears, and it is evident the beast intends to devour them. Watonai engages the animal, but is unable to subdue it until his mother comes to his aid and produces a lucky charm which she brought with her from the Grand Shrine at Ise, in Mie Prefecture, Japan.

Left: Actor, Arashi Rikan, II (1788-1837) as Watonai struggles with a tiger, from the c.1833 *Ukiyo-e* print by Ganjosai Kunihiro, (1815-1843)

Right: *Watonai Riding a Charmed Tiger* from the *Ukiyo-e* print by the artist Utagawa Kunisada (1786-1865)

The American born Japanese scholar and "Japanologist", Donald Keene translates what happens next from Chikamatsu's text: "Watonai!" cries the warrior's mother. "You were born in the Land of the Gods [Japan], and you mustn't harm the body, hair and skin you received from them in a contest with a brute beast. Japan is far away, but the gods dwell in your body. Why shouldn't this charm from the Great Shrine by the Isuzu River be effective now?" After taking but a moment to consider his mother's proposal, Watonai grabs hold of the proffered charm and points it toward the tiger who is subdued magically by the power of the amulet.

In an ensuing encounter against a substantial force of Chinese troops, Watonai screams at the Chinese leader, "You obviously despise Japan as a small country, but are you not impressed with the martial art of the Japanese, which even this tiger fears?" In the next moment, the enemy *Han* turn their attention in an attempt to overcome Coxinga's mother. Keene tells us that the Chinese instead, "...make a beeline for her with flailing swords, but in a further sign of his divine protection, the gods lend their strength to the tiger. It springs up, quivering, and bares its teeth. It leaps with a fierce roar at the enemy...They fling at the tiger their hunting spears, rough lances, and whatever else comes to hand, and slash with their swords. The tiger, possessed of divine strength, leaps about at will, snatching their swords in mid-air with his jaws, and dashing them to splinters against the rocks." With the tiger there for support, Watonai triumphs over the Chinese. In the end, the tiger demonstrates that it is an aid to all of those future Japanese soldiers who call upon him for victory.

Actor, Ichikawa Danjuro, IX (1838-1903) as Watonai in the c.1898 *Ukiyo-e* print by Toyohara Kunichika, (1835-1900). (Note the *hariko-no-tora*, or bobbing head tiger doll in the upper right)

The actor Kawarazaki Gonjuro 1 as the historical figure Watonai from the c.1863 woodblock print by the artist Toyokuni 3/Kunisada (1786-1864)

Kato Kiyomasa: The Conqueror of Korea

One of the more popular tiger hunting *musha-e* legends, focuses on the historical samurai warrior, Kato Kiyomasa (1562-1611). Known by various names including Toranosuke and "The Conqueror of Korea", Kiyomasa's fabled actions against the powerful orange and black striped beasts, inspired some of his admiring contemporaries to test their mettal in similar fashion in order to see how they measured up against the prowess of their hero.

Kato Kiyomasa was born in what is now Nakamura-Ward, near Nagoya City, Japan. His mother was a cousin of the famous samurai warrior, and daimyo, Toyotomi Hideyoshi. Early in his career, he fought in Hideyoshi's army at the Battles of Yamazaki (1582) and Shizugatake (1583). His commanders noted that he was both courageous, and ruthless in combat.

By their merit in these battles, Kato Kiyomasa and six other attending samurai warriors became known as "The Seven Spears of Shizugatake". Along with this prominent title (which may actually have been a later Edo era construct), a couple of the men were given large land holdings and some became *daimyo*.

The Great Battle of Yamazaki, Ukiyo-e triptych, c.1865 by the artist Tsukioka Yoshitoshi (1839-1892).

The Seven Spears of Shizugatake, woodblock triptych by Utagawa Toyonobu (1859-1886), c.1883.

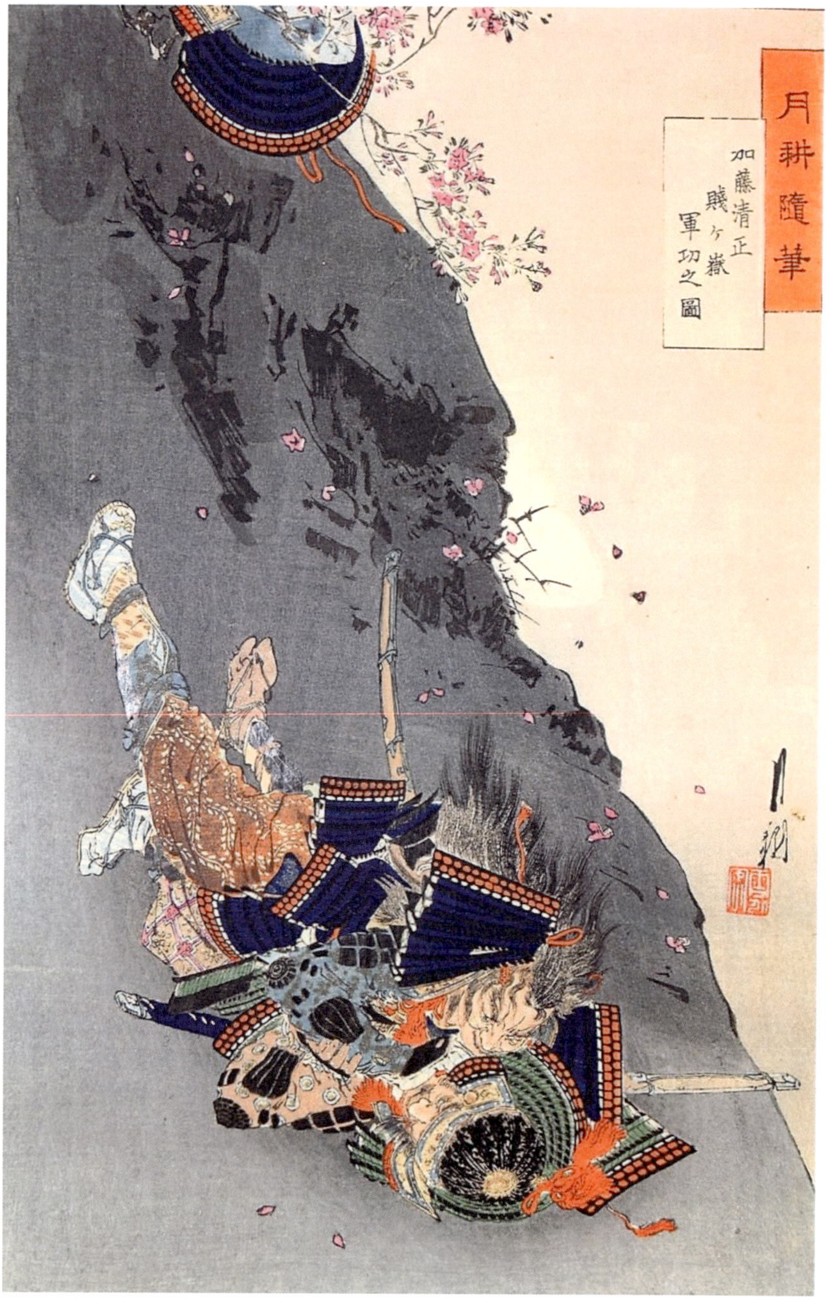

Kato Kiyomasa Gaining Meritorious War Service at Shizugatake, from the c.1896 *Ukiyo-e* print by Ogata Gekko (1859-1920)

Later, as one of three senior military commanders during the Seven Year War (1592-1598), Kiyomasa led men against the Korean Dynastic rulers of Joseon, initially founded by Yi Seong-gye in 1392. His bold military leadership and tenacity earned him the nickname "The Demon", or "Phantom General", from the Koreans, and enabled his troops to defeat their enemies defending Seoul, and Busan. His last major engagement at the Battle of Imjin River was instrumental in bringing organized resistance to an end.

During the Conquest of Korea, Kato Kiyomasa, and other samurai were known to participate in the "sport" of hunting tigers; his weapon of choice was the spear. According to legend, once the animals were dispatched, Kiyomasa ordered that the pelts were to be presented to Toyotomi Hideyoshi. Different versions of this story have emerged over the years: one anecdotally refers to the intent of capturing the tigers alive so that they could be caged and shipped back to Japan.

It is said that Kiyomasa planned to provide Hideyoshi with fresh tiger meat as a health aid (tiger body parts are still believed to have health and healing properties today in parts of Asia, much to the tiger's obvious detriment). The captured beasts were killed, however, before they could be sent home, having been fed to the hungry Japanese soldiers in the field. Interestingly, during the Korean Campaign, it was Hideyoshi who would later ban his officers from taking part in the sport of hunting tigers. Apparently too many men were being killed in the attempt to compete with their idol, Kiyomasa.

Kato Kiyomasa in Korea Viewing Mt. Fuji Under a Full Moon. Woodblock print (1893), by the artist Nobukazu Watanabe (1874-1944).

Kato Kiyomasa on the Korean Campaign from the c.1885 *Ukiyo-e* triptych by Koyama Chikusai, (Meiji era: 1868-1912).

5. Bobbing Head Tigers

"The most difficult thing is the decision to act, the rest is merely tenacity. The fears are paper tigers. You can do anything you decide to do. You can act to change and control your life; and the procedure , the process is its own reward." – *Amelia Earhart*

A History Of The Bobbing Head Tiger Papier-Maché Doll *(Hariko-no-tora)*

The history of the bobbing head tiger in Japanese culture is multi-faceted and spans hundreds of years. The story begins with the formation of the zodiac calendar and the 12 animals that occupy it. Shrouded in lore, there are a number of popular tales that surround it. According to one Buddhist legend, knowing that he would soon depart the earth, the Buddha called before him all of the animals of the earth; only 12 animals arrived to pay their respects. To show his gratitude to them, each animal was rewarded by having a year named in his honor according to the order in which they appeared.

As the third animal of the Asian zodiac, the tiger is said to have numerous qualities. For the ancient samurai as well as his World War Two era contemporary, the tiger was seen as sensitive but also stubborn and resolute. Tigers have enormous courage, and project a "can do" personality that is much admired whenever difficult circumstances arise, making its image a meaningful subject to be added to any good luck flag.

The tiger *ningyo* or "dolls" illustrating this style of good luck flag are emblematic of the Miharu Hariko papier-maché dolls that have been famously produced in Japan since the Edo period (1600-1868). Miharu-*machi* or Miharu town is located in a mostly rural part of Fukushima Prefecture. The area has a long history of folk crafts being produced there. *Mi-haru* or "3 spring" is so named because of the cherry, plum and peach trees that bloom heavily throughout the immediate countryside.

During the *Sengoku* period (1467-1603), Japan experienced almost continuous civil and military upheaval. The warring finally came to an end with the establishment of the Tokugawa shogunate. Under Tokugawa rule, the period known as *sakoku,* or national isolation, was established. No foreigners were allowed to enter Japan, and no Japanese were allowed to leave the country, under penalty of death.

It was during this time of unrest that Hashimoto Keibu, a minor samurai warrior, settled in the area. His descendants were the craftspeople that first made *Miharu ningyo* under the sponsorship of local feudal lords. The process of manufacturing *papier-mâché* was believed to have been exported to Japan by the Chinese sometime during the *Muromachi* period (1333-1568). A lack of clay in the area made it impossible to make dolls similar to the well known ones manufactured in Sendai-han. Therefore, *Miharu ningyo* were made from scraps of dampened mulberry paper pressed over willow wood molds. Once they dried, the shapes were glued together using cow skin glue, then covered with *gobun,* or crushed shell powder-paint. After drying, the figures were colorfully painted.

The popularity of the little dolls grew during the period of the Shogun's *san kin kotai* system. Under this practice, important government officials and high ranking samurai were required to divide a portion of their time living between their outlying homes, and the capital city of Edo (Tokyo). It was recognized that by spending time between both locations, the samurai would find it difficult to unite and plot any further uprisings. *(See appendix 3, page 192).*

The major thoroughfares that intersected Edo were thereafter heavily traveled by the samurai, and their various retinues. The practice of purchasing souvenirs to take back home grew to its zenith sometime in the 1700s when numerous families were employed in all aspects of making *ningyo.* Eventually a number of different towns with proximity to the key ingredients: paper manufacturing and cold fresh water began to make *papier-mâché* dolls, although Miharu remained prominent. Descendants of the original family continue to make bobbing head tiger dolls to this day.

The larger *papier-mâché* tiger doll pictured on page 37, with its bottom raised, head turned toward the left and mouth open in a growl, follows a tradition of *hariko,* or folk dolls, that date to the Edo period. The position of the tiger identifies it as a male tiger. There is a corresponding female doll that has its head turned toward the right-hand side, with its mouth closed. The two together represent *Kongorikishi,* or the "*A-Um*" spirit of Buddhism.

Kongorikishi are two powerful guardians of the Buddha, and are manifested as statues positioned at the entrance of many East Asian Buddhist temples. According to Japanese tradition, their frightening countenance was due to their job of protecting Gautama Buddha during his travels. While Buddhist tradition typically exemplifies pacifism, there are exceptions to this as when defending enshrined values and ideals against the forces of evil.

The Tokaido March by the extremely popular artist, Utagawa Kunisada (1786-1865). This woodblock print portrays the samurai, and their retinue crossing a river on their way between Tokaido, and Edo. The lack of bridges and fast moving water made it necessary for travelers to be ferried by boats, or carried bodily by watermen.

Examples of tiger *ningyo* or dolls.

This guardian pair of figures are normally positioned within a separate temple entrance gate. The right-hand statue is identified as Misshaku Kongo, and has his mouth open, symbolically vocalizing the first part of the Sanskrit, "*A-*". The corresponding statue on the left, known as Naraen Kongo, normally has its mouth closed, vocalizing the second aspect of the Sanskrit, "*-Um*". When the two are combined, they form the sound, "*A-Um*", which means, "The Absolute" in Sanskrit. It is symbolic of the birth, death, and re-birth that all sentient beings endure. The larger tiger doll shown above was a common gift following the end of World War Two.

The smaller *papier-mâché* tiger dolls shown above have a number of different names: they are sometimes called *Yokomuki Tora*. In English, the phrase translates as "Tiger Looking To The Side". Examples generally have a movable head that is secured to the inside of the tiger's body that allow it to pivot to either the left or right. Similar looking tiger images or figures of this sort may be called *Mayoke Tora*, which translated simply means "Tiger Amulet/Tiger Warding Off Evil". Either the small charm, or an image of it, was believed to help the bearer avoid evil influences and is popularly used to keep children safe and healthy.

As shown, *papier-mâché* tigers came to be carried as talismans of good health and protection. This was especially true during and after the first great Asian cholera pandemic (1817-1824). Cholera was endemic to parts of Asia particularly along sections of the Ganges delta where populations had experienced the effects of the disease over the years. Some believe that Hindu pilgrims contracted and then spread it in 1817 following the traditional festival known as *Kumbh Mela*. The contagion was eventually exported to other sections of India and by 1822 made its way to Japan and beyond by traders, probably via the trading port at Nagasaki.

People died within a few days following exposure to the cholera organism. Many in Japan became so frightened from the symptoms and effects, that they coined a phrase for the illness, calling it "The Sickness of the Tiger and the Wolf". Some said that suffering from the disease was as if the person had been attacked by both ferocious animals at the same time.

In those days, nothing was known with regard to deadly microorganisms and their affects upon people's health. A popular treatment at the time involved ingesting a combination of herbs and powdered tiger bone. Thereafter, people began to associate tigers with having great power to protect and heal. *Papier-mâché* tiger dolls were then sold as charms for good health and luck. Over time, their images, drawn or painted upon objects such as a *hinomaru yosegaki* came to represent objects of good fortune, intended to keep away evil and inspire the bearer to be filled with courage.

Bobbing Head Tiger Flag - 27" x 41" (68.58cm x 104.14cm).

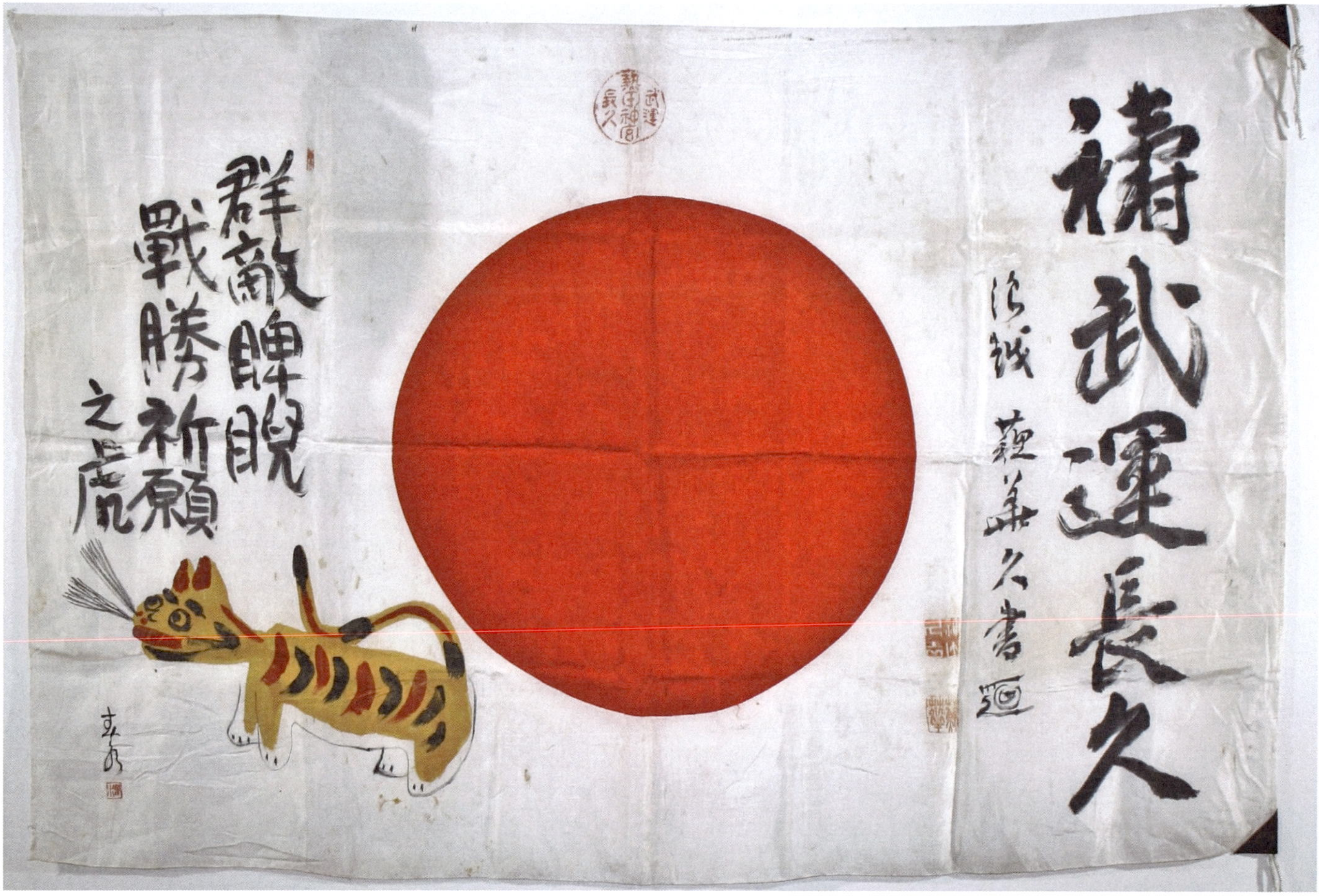

This good luck signed flag is made from a silk material that has darkened only slightly with age. Both brown leather corner tabs are sewn to the hoist ends of the flag, and each has its white tie strings firmly attached. The hoist and fly edges of the flag have stitches added to strengthen those areas. The red sun in the center has been screen or block pressed onto the white field.

The large dark vertical row of characters closest to the hoist, or right-hand edge, say *Ki Buun Chokyu,* or "I Pray Your Military Fortunes Are Long Lasting". The smaller row of vertical characters just to the left of those say "Writing By Numakoshi Sokahisa". The two square seals found to the left of that are illegible.

Detail of Bobbing Head Tiger, 7.25" x 14.75" (18.42cm x 37.47cm).

In the top center of the flag near the 12 o'clock position, a large round seal may be seen. That seal measures 2.5" across, and says *Buun Chokyu,* or "May Your Military Fortunes Be Long Lasting". The remaining characters within the seal say "*Atsuta* Shrine". The *Atsuta* Shrine is a *Shinto* shrine located in Nogoya, Japan. It is believed to have been established by the Emperor Keiko between the years 71-130. Of important note is that the *Atsuta* Shrine houses the *Kusanogi no Tsuruge*, an ancient and highly revered sword known as one of the three Sacred Treasures. The other two treasures are the mirror and the jewel. Ancient *Shinto* belief reveals that the sword was a gift from Amaterasu Omikami, known as the Sun Goddess from whom the Japanese Emperors were said to be directly descended.

On the left-hand side of the red sun are large vertical lines of ideograms. These combine to say "Glare Down At the Swarming Enemy – Battle Victory Tiger!" The black, white, orange and red painted tiger illustrated below those characters represents an artistic style seen often in Japan that is popularized by having a bobbing head. This form of tiger is widely collected, and may be seen carved from wood or made from other materials, including papier maché. It is also portrayed extensively in toys, paintings, decorative cards and other forms of art and advertising. The artist's signature and seal may be seen below the tiger's chin. The signature reads Shunsui or Harumizu.

Bobbing Head Tiger Flag – 26.5" x 33.5" (67.31cm x 85.09cm).

The colorful silk good luck flag pictured above is in very nice overall condition. Both gold foil paper corner tabs are in place as are their white cotton tie strings. The fly and hoist side edges of the flag have added stitching to provide strength to these high stress areas. The red center has been screen pressed onto the white field. Of note is the fact that this flag was signed with the hoist side edge on the left-hand side rather than on the right. This is simply a variation as most flags encountered are signed with the corner tabs/tie strings on the right-hand side.

The first row of four vertical characters nearest the left-hand edge are a name, Ueda Takehiko. The large vertical row of five ideograms to the immediate right of that are for the slogan *Ki Buun Chokyu,* or "I Pray Your Military Fortunes Are Long Lasting". Just below the lower character for *Ki Buun Chokyu* are two thinly written stylized vertical lines of characters. In Japanese these say, *Kimi no Tame Nanika Oshiman Wakazukura, Chitte Kai Aru Inochi Nariseba.* A literal translation of this reads "Young Cherry Blossoms Have Nothing To Regret As It Is Worth Giving Up One's Life For The Sake Of The Emperor". There are quite a few slogans written upon this flag. The large horizontal row of four characters near the top edge in the 11-12 o'clock position in front of the tiger's face say "Serving The Nation With Utmost Loyalty".

In the upper right-hand corner, the large vertical row of characters next to the black rounded figure of *Daruma* say, *Ki Gokento,* or "Prayers For A Good Fight". Immediately below that slogan is another vertical row of four large characters. Those read, *Ken Teki Hissatsu,* or "Search And Destroy". Below the rounded figure are two dark, medium-sized characters that say *Hissho* or "Certain Victory!" Dropping below this and slightly to the right are four dark, medium-sized vertical characters. This slogan reads *Seki Shin Hokoku,* or "Serving The Nation With A Pure Heart".

The bobbing head tiger illustrated in the center/upper right-hand corner was painted using yellow, black, blue, white and red paint. He is portrayed with his snarling mouth open and teeth barred. A red artist's seal that measures 0.5" (1.27cm) square may be seen just below the tiger's chin. While highly stylized, the name on the seal appears to say Toride. The three black ink *kanji* to the right of that give another name, Morinosuke. Writers, poets and artists often took pen names that were not their actual ones, and this may be the case with the two examples here.

Detail of Bobbing Head Tiger, 4" x 6.5" (10.16cm x 16.51cm).

To the right of the tiger, a large rounded figure of Daruma may be seen. Known also as the *Bodhidharma* or simply *Dharma*, he was a Buddhist monk who lived during the 5th or 6th century. Details of his early life vary, however, he appears to have come from either central Asia or the Indian sub-continent, and was given credit for exporting Ch'an or Zen Buddhism to China. The Japanese celebrate the image of Daruma as a symbol of good luck, and fortune. For that reason, it may be seen drawn or painted upon good luck flags from time to time.

Stories surrounding the figure and image of Daruma are plentiful. One of the more standard ones says that his figure was made rounded so that whenever it is knocked down it will quickly roll back upright. This is to emulate his persistence in seeking enlightenment. Tenacity in any form is an admirable quality, embraced by the Japanese serviceman.

During the war, pictures or statues of the deity were common gifts. These new tokens of luck generally had blank eye holes. When a specific goal or wish was determined, prior to the man's departure overseas, one of the eyes might be filled in upon the face, while the second could be added following a safe return home. *Daruma* figures painted upon good luck flags with only one eye are sometimes said to show that the owner did not return home from war. In some instances the story is likely true.

Bobbing Head Tiger Flag – 27.5" x 38.75" (69.85cm x 98.43cm).

Detail of tiger, 3" x 5" (7.62cm x 12.70cm).

This slightly longer than standard silk good luck flag (above) is in near excellent condition with no stains or holes present. The heavy white paper corner tabs are present and sewn into place. They have been pressed in order to give the paper a rough leather appearance. There are silk tie strings present on both corners. The red sun was screen or block printed onto the white background.

Prior to signing, the flag was placed in the more usual fashion with the corner tabs on the right-hand side. Again, while most flags are seen positioned in this manner, numerous others were placed with their tabs on the left-hand side before they had signatures added. Placed horizontally across the top of the flag are five large, prominent characters that say *Ki Buun Chokyu,* or "I Pray That Your Military Fortunes Are Long Lasting".

The large, yet slightly smaller vertical characters along the right-hand side give the name of the flag's owner "Presented to Koyama Tsuneo". A black and white tiger caricature is seen near the red orb in the 7-8 o'clock position. The long lines of characters above it encourage "Go And Become The Cornerstone For The Rise Of Asia". The flag's white field contains nearly 50 signatures, however, only one seems to list that person's title, "Navy Medical Captain Okuda Ryuichi".

Sitting for a photograph at his desk, this Japanese sailor displays a drawing of a bobbing head tiger on the wall behind him, perhaps a gift from a loved one.

6. Tiger Head Portraits

"The focused mind can pierce through stone." – *Japanese Maxim*

Tiger Head Portrait Flag - 26" x 36.5" (66.04cm x 92.71cm).

The good luck signed flag pictured above is made from a fine silk and is in wonderful condition. Each of the hoist side, white paper corner tabs are sewn into place and both maintain their silk tie strings. The hoist and fly side edges of the flag have additional stitching added for strength. The red center orb is silk screened onto the white field.

In the upper right-hand corner of the flag, a colorful growling tiger's head measuring 3.5" long was painted using shades of black, white, red and orange ink. Located directly above it, a small black and white shrimp (*ebi*) was also drawn. Shrimp have long, curving antennae, and their swayed backs are similar in appearance to those of the elderly. In Japanese culture, this makes them a symbol for longevity.

In addition, because shrimp molt their shells, they have become emblematic as a symbol for the renewal of life. The artist's name may be seen just below the shrimp but it is difficult to read. The first horizontal line of characters written just above the tiger's head read *Giyu Hoko*, or "Serving With Justice And Bravery". The two ideograms just to the left front of the tiger's mouth say *Sei Hei*, or "Sacred Soldier".

The four largest horizontal *kanji* characters written across the top say *Buun Chokyu*, or "May Your Military Fortunes Be Long Lasting". The largest vertical line of characters located nearest the fly edge say *Shuku Nyuei* Tada Kyuhachi-kun, or "Celebrating Induction For Mr. Tada Kyuhachi". Just to the left of this line, the large vertical line of characters reads *Tana Butai Cho* Kurokawa *Shosa*, or "Chief of *Tana* Unit, Major Kurokawa". The *Tana* Unit was a branch of the military arms arsenal located in Kanagawa prefecture in the southern Kanto region of Japan. The capital of the prefecture is Yokohama.

Immediately to the left of these two larger vertical lines are a few small dark lines of vertical *kanji*. Starting with the uppermost, the first says *Sei Ken*, or "Sacred Progress"; the next says *Sei Shi*, or "Sacred Death"; *Sei Sho*, or "Sacred Victory" follows that, and the final ideograms say *Sei Sen*, or "Sacred War".

Detail of tiger head portrait measuring 3" x 2.5" (7.62cm x 6.35cm). The shrimp measures 2" x 2" (5.08cm x 5.08cm).

In the right-hand corner, just below the two largest vertical lines of characters, are an additional two lines of dark horizontal *kanji*. These lines are actually positioned on their side and should be read vertically. Those lines say *Sichisei Hokoku,* or "Seven Lives In Devotion To The Nation" and are followed by the name Mitsui Katsuichi.

The *samurai* warrior Kusunoki Masashige and his brother Masasue became legendary during the *Genko* War period (1331-1333). In service to the Emperor Go-Daigo, both men and most of their clansmen would eventually be badly defeated in 1336 at the Battle of Minatogawa. During the engagement, Masashige's brother was said to have uttered this now famous phrase, *Sichisei Hokoku,* or "Would That I Have Seven Lives To Give To My Emperor Prior To Death." Masashige also lost his life during the same action that took place against the *Ashikaga* clan.

During the Second World War, members of the special attack units, or *tokkotai,* as well as ordinary Japanese military and citizens saw themselves similarly, as they willingly offered their lives up for the Emperor/country.

Written near the edge of the red sun in the 1 o'clock position are small, neat characters left by a female admirer by the name of Kaneko Kuniko. They read "Mr. Tada Kyuhachi - Do Your Best And Decimate The Americans And The British!" In the upper left-hand corner are two dark, prominent characters (one looks like a house with a curved roof on it). Those give the name Kanehiro.

Kusunoki Masashige arrives at the Go-Daigo's *anzaisho* to pay his respects.

Tiger Head Portrait Flag - 27″ x 33″ (68.58cm x 85.73cm)

This good luck flag is made from a heavier silk material that has age toned and is now quite dark in color. Both leather corner tabs remain firmly in place and each has its silk tie strings remaining. The hoist and fly side edges have additional stitches placed by the manufacture that provides extra strength. The red center orb, or sun, has been silk screened onto the white field.

The flag is covered with numerous signatures and slogans, and has a tiger's head reminiscent of "Tony the Tiger" (trademark of the Kellogg's Company), near the upper left-hand corner at the 10 o'clock position. The two largest characters drawn above the orb in the 12 and 1 o'clock positions say *Seishen,* or "Faithful". The largest character grouping in the upper left-hand corner say *Kenyu,* or "Strictly Courageous".

The tiger head measures 3.5" x 2.75" (8.89cm x 6.99cm).

Tiger Head Portrait Flag – 27.75" X 33.25" (70.48cm x 84.45cm)

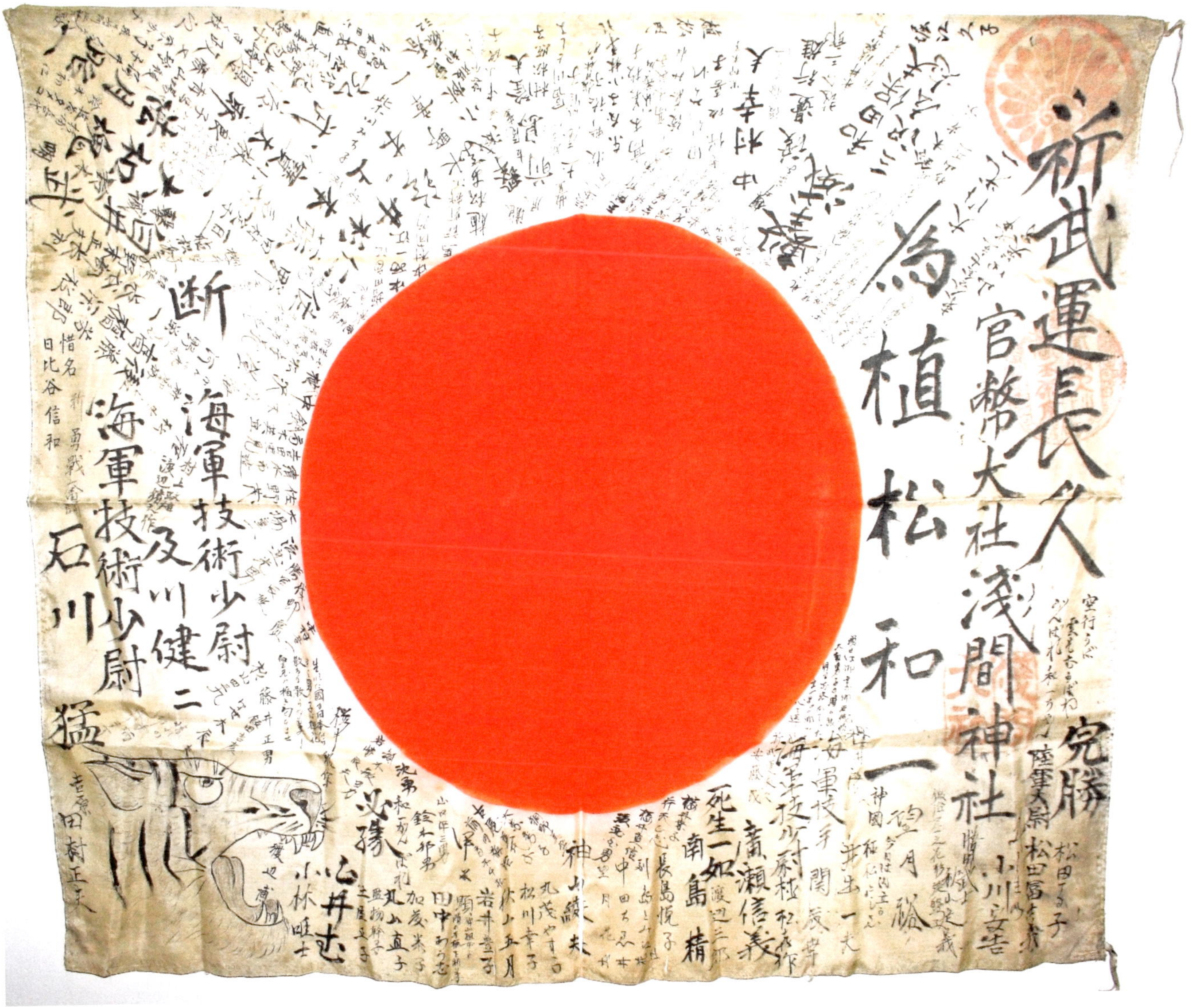

The Imperial *Asama* Shrine seal.

This marvelous Japanese good luck signed flag is made from an age toned white silk. The hoist, in this case the right-hand side of the flag, has two corner reinforcing tabs made from separate pieces of white silk that were sewn to the reverse corners. Flat, white silk string ties were then securely attached to each corner by added stitches. The hoist and fly sides of the flag were given extra stitches in order to provide added strength. The faded orange-red orb was screen printed into the center of the white field.

The flag displayed here is profusely signed. In addition, there are numerous poems, song lyrics, exultations and beautiful red seals, included to remind the owner to perform his duty. Near the flag's hoist side edge are five bold vertical *kanji* characters that say *Ki Buun Chokyu,* or "I Pray Your Military Fortunes Are Long Lasting". To the left of that, another longer line of vertical characters say *Kanpei Taisha Asama Jinja,* or "Major Imperial Shrine – *Asama* Shrine". That line of characters was applied over a large, red square seal that measures 3" (7.62cm) and reads *Asama Taisha.*

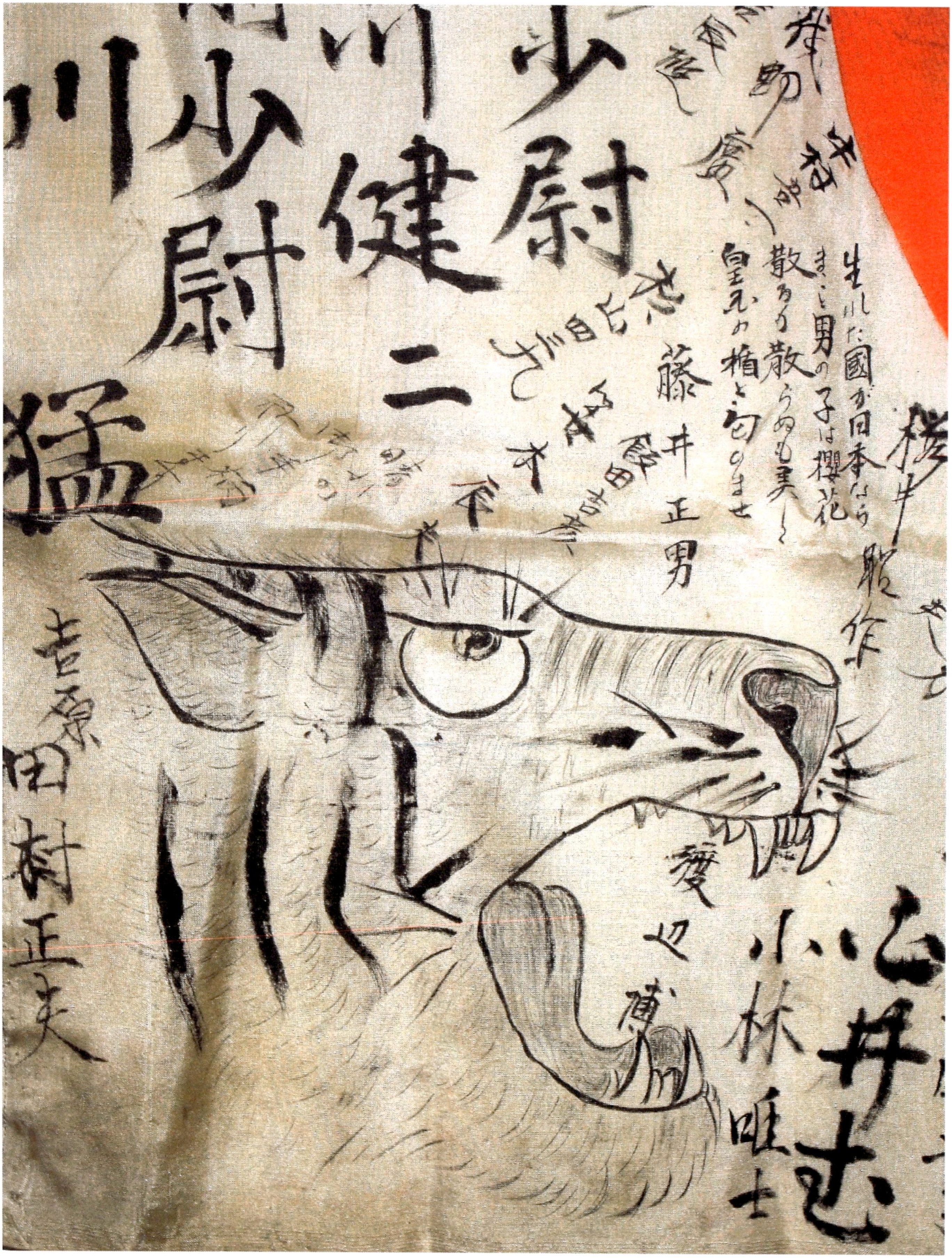

The tiger head portrait measures: 7.25" x 7" (18.41cm x 17.78cm)

In their fundamental form, the *Asama* Shrines are particular *Shinto* Shrines that embody the worship of *kami,* or the gods of mountainous volcanoes, especially Mount Fuji. As Buddhism's popularity grew and it acquired adherents throughout Japan, many of the popular *Shinto* deities were superseded by their Buddhist versions. The *Asama* Shrines support belief in the legendary mythical princess Konohanasakuyahime. Not only is she the deity of Mount Fuji but she is also recognized as the deity of blossoming cherry trees.

In *Shinto* mythology, Konohanasakuyahime, known too as Konohana, is the daughter of Oyamatsumi, the god of mountains. According to early lore, she married the Sun Goddess Amaterasu's grandson, Ninigi-no-Mikoto who would later become the great-grandfather of the legendary Emperor Jimmu. In a single night, she became pregnant, and while covered in flames, gave birth to three children. Her final fate is mixed: some accounts saying that she survived the fiery ordeal, while others say that she died. From the latter, Konohana therefore is likened to the fabled cherry blossoms that fall at the height of their beauty. Today, hundreds of *Asama Shinto* shrines exist in Japan, although the most influential are those that approximate the summit of Mount Fuji.

Japanese military troops training in the mountains that surround Mount Fuji. The man on the left-hand side of the image holds in his gloved hands the unit's tattered, yet revered battle standard.

The following vertical line of characters immediately to the left of the *Asama jinja* seal say, *Tame* Uematsu Kazuichi, or "For Uematsu Kazuichi", the flag's owner. Moving from that location clockwise toward the 7-8 o'clock positions along the edge of the red sun, is a block of four vertical lines of *kanji* characters that represent a poem. Translated into English they say "If [he] Is Born In Japan, That Boy Is Truly A Cherry Blossom, Whether He Dies Or Not, Always Being The Shield For The Emperor. [Signed by] Fujii Masao"

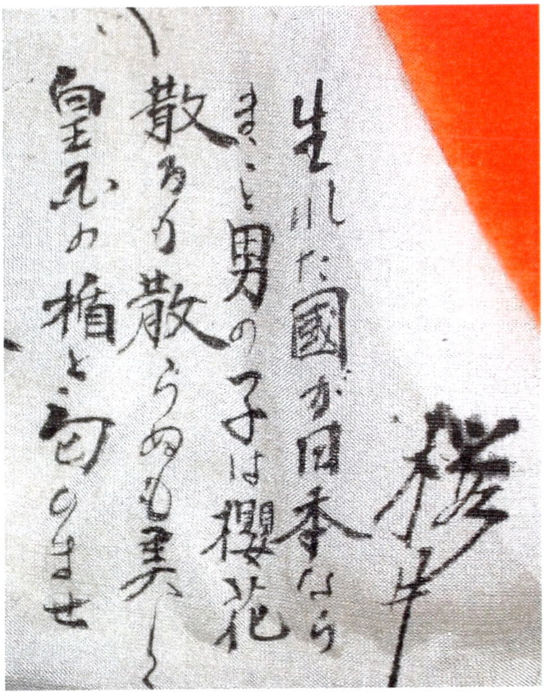

To the left of the poem are another four vertical lines of dark, bold ideograms; reading them from right to left they identify two friends of the flag's owner. They say, *Kaigun Gijutsu Shoi,* Oikawa Kenji, or "Navy Technical Second Lieutenant Kenji Oikawa" and *Kaigun Gijutsu Shoi,* Ishikawa Takeru, or "Navy Technical Second Lieutenant Takeru Ishikawa".

In the upper left-hand corner, located between the 10-11 o'clock positions, is a lyric from the well known wartime patriotic song/poem, *Umi Yukaba.* This specific patriotic song, or *gunka,* is based upon the style of poem known as a waka poem, found in the *Man'Yoshu* with lyrics by Otomo no Yakamochi (718-785). Yakamochi was one of the most famous and active poets of his day, helping to compile Japan's first collected works of poetry. The poem was set to music by Kiyoshi Nobutoki in 1937, and was called Japan's second national anthem during the war. Popular to the very end of the conflict, *Umi Yukaba* was played often and sung by some members of Japan's special attack squadrons as they were sent off on their final missions. *(See appendix 4, page 192).*

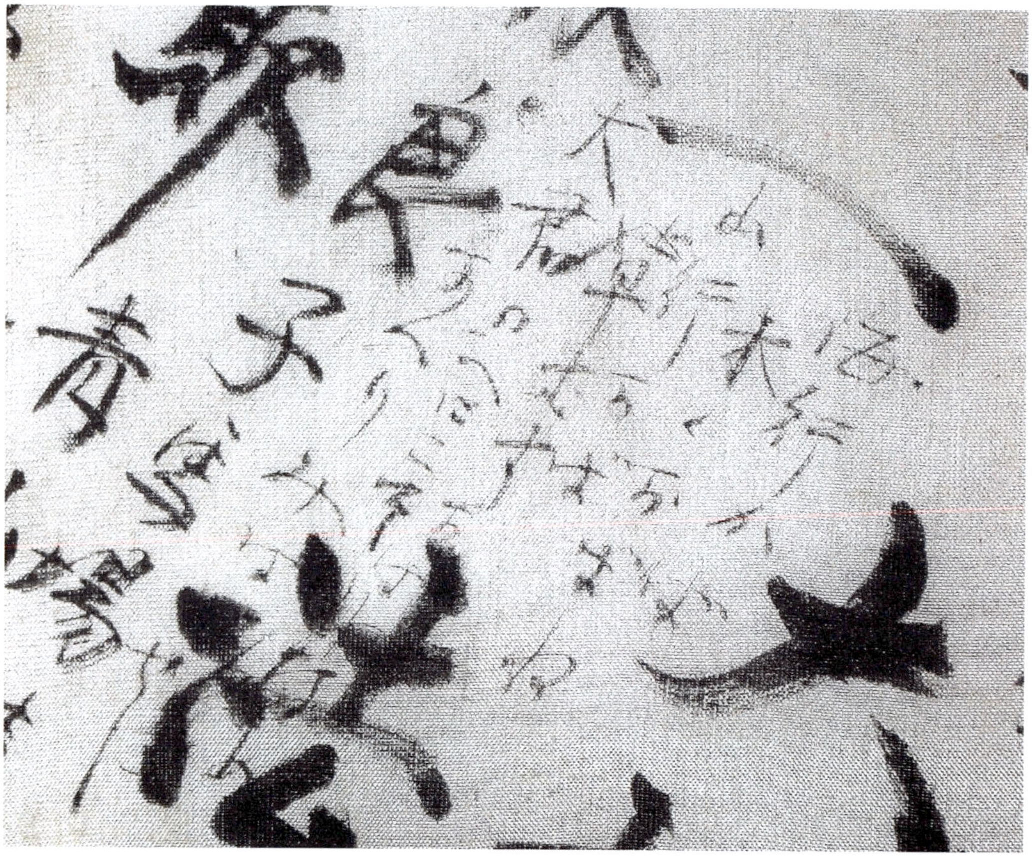

A portion of the song lyric from the *Umi Yukaba* by Otomo no Yakamochi.

Continuing clockwise from the uppermost left-hand corner toward the upper right-hand one is an emotional message.

This appears to be a declaration from a man, possibly the flag's owner, to his sweetheart, Reiko. It reads "When the time comes for me to leave for battle, please do not cry my sweet. Just see me go in grace, that is the spirit of the man, always keep in mind. The memories of our all night talk... shall not forget our memories. April 1945, the day we were up all night."

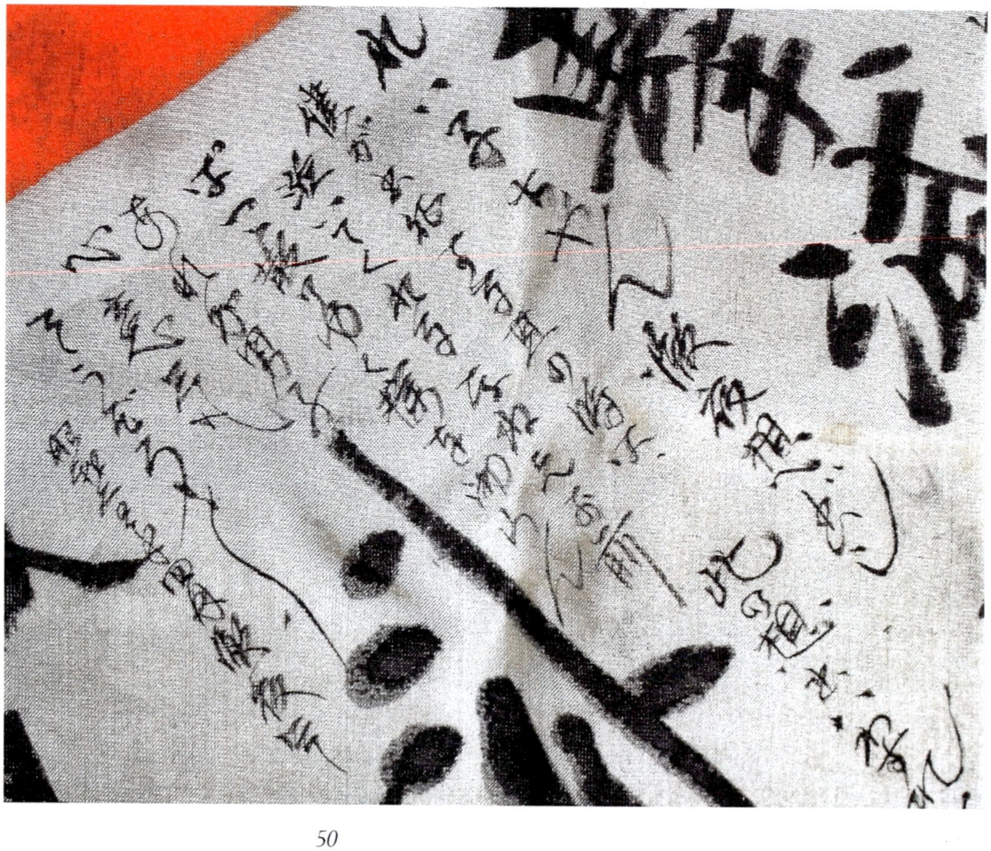

In the lower right-hand corner near the hoist side edge in the 4 o'clock position, is a series of *kanji* characters that were added later to the *Umi Yukaba*, and it pertains to pilots. The Japanese reads *Sora Yukaba, Kumo Somu Kabana* or "When One Goes To The Skies, His Corpse Shall Dye The Clouds". This is followed by the admonition *Gambare* Kazuichi-Chan, or "Do Your Best Kazuichi-Chan".

It is most probable that this good luck flag belonged to a navy student pilot preparing for the *Shinpu Tokubetsu Kogeki Tai,* or "Divine Wind Special Attack Units". The flag, accompanying photographs, as well as other documents and items related to this pilot's navy career were all found together. The exact flag pictured can be seen in the following individual images being held by its owner.

The first and second images show Kazuichi Uematsu, along with friends and family, displaying his *hinomaru yosegaki* for the photographer. The tiger head image clearly stands out in the lower left-hand corner of the flag. The full and quite distinct round red seal may also be seen in the upper right-hand corner; the vertical line of characters for *Ki Buun Chokyu* may be viewed written over the top of it, along the right-hand margin. The large square seal located just below the 3 o'clock position is also evident in the photograph.

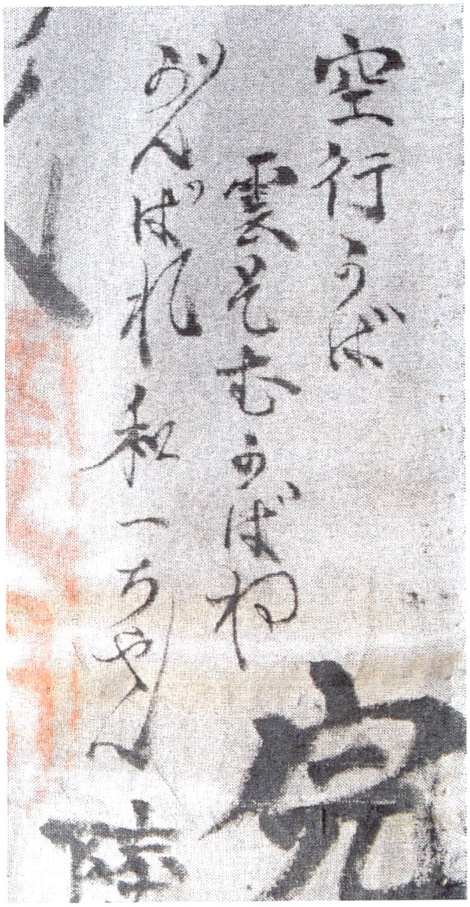

A portion of the *Umi Yukaba* for pilots

Kazuichi Uematsu (front row, 6th from the left) holds his tiger illustrated flag.

The tiger head illustration is visible in the lower left-hand corner of this detail from the photograph on the previous page. The distinct round seal is also evident in the upper right-hand corner.

The reverse side of the photographs provide interesting information with regard to the history of Uematsu Kazuichi at war's end. One inscription says, "Uematsu Kazuichi prepares to leave for induction in The Koriyama Kaigun Kokutai in Fukushima." A second line of text says, "Uematsu Kazuichi, ready to go with flag 15 minutes before leaving on May 25, 1945."

The Koriyama Naval Air Group was initially formed on February 1, 1944 but was dissolved about a year later. Per Notice Number 127, Naval Air Group Organizations Amendment, Minister's Secretariat, Ministry of the Navy dated February 11, 1945 a new unit was formed from the old organization, and officially renamed the 1st Koriyama Naval Air Group on March 15, 1945.

Uematsu Kazuichi prepares to leave home on May 25, 1945; one hundred days later, the war was over. Note: the good luck charm pouch that hangs from his belt loop. The photograph shows Uematsu Kazuichi wearing a *hachimaki* (headband).

Detail of the headband. The black *kanji* characters on either side of the *hinomaru*, or sun, in the center read *kamikaze* or "Divine Wind".

Tiger Head Portrait Flag – 27.25" x 36.75" (69.21cm x 93.34cm).

This is a simple, yet elegant Japanese good luck signed flag with tiger art image. The flag is made from a white silk material that is only slightly stained. The tabs, set in the right-hand corners of the flag, are made from a gold foil paper. Both tabs no longer have their tie strings. Typically, the fly and hoist sides of the flag have been re-enforced with added stitches to provide strength to those sides of the flag. The red *hinomaru* or red sun has been silk screened onto the white field.

The bold, dark vertical line of *kanji* characters closest to the hoist, or right-hand side, of the flag say *Shinshu Kenji Iki Chuuten,* or "The Stalwart Youth Of God's Country Has A Spirit That Reaches The Heavens". The long vertical line of characters to the left of that says *Rikugun Gijutsu Shosa* Nakai Kagami, or "Army Technical Major Kagami Nakai". The last bold line of vertical *kanji* to the left of that say *Kougeki Seishin,* or "Attack Spirit!"

The two large characters at the 12 o'clock position of the flag say *Toryu,* or "The Climbing Dragon". Immediately below that are four smaller characters written in a horizontal position. Those characters give the name "Technician Inoue".

Just above the tiger's neck are two small red seals that each measure 0.5" (1.27cm) square. They are difficult to read, as are the black *kanji* characters placed along side them in ink. However, one of the stamps gives the name Iwamoto, which is probably the name of the person who painted the beautiful tiger image.

Tiger head portrait measures 8.5" x 6.75" (21.59cm x 17.14cm).

Tiger Head Portrait Flag – 26.75" x 31.25" (67.94cm x 79.37cm). *Flag Courtesy of Vintage Productions*

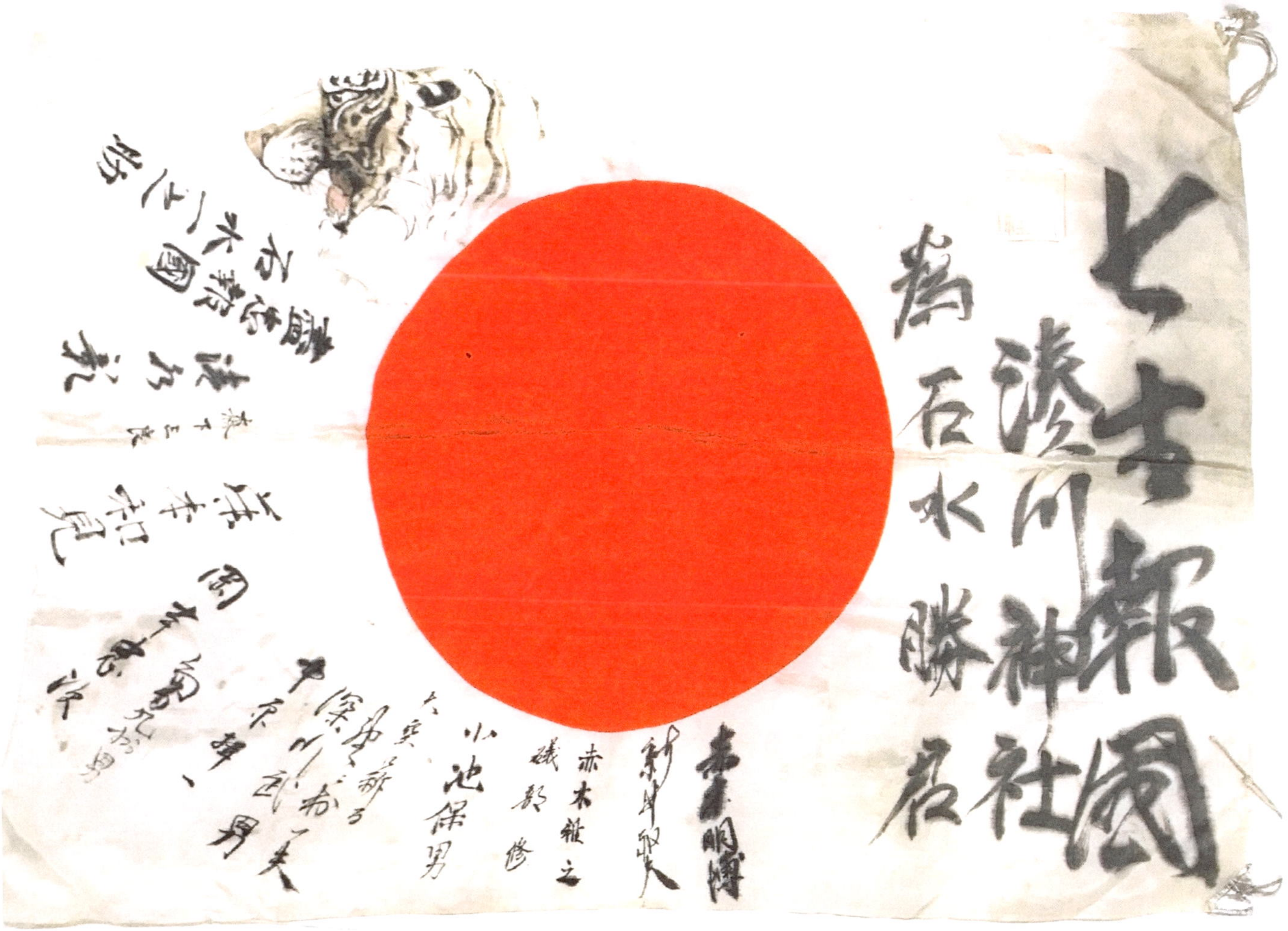

This Imperial Japanese good luck signed flag is made from a slightly darkened white silk. The hoist and fly side edges have the usual hem stitching to provide strength, while the silver paper foil corner tabs, with tie strings remain in place. The deep red colored sun was silk screen pressed onto the field's white background.

On the flag's right-hand side appear 3 large vertical lines of bold *kanji* characters. The example nearest to the hoist side edge say *Hichisei Hokoku,* or "Seven Lives In Devotion To The Country". The center line of ideograms spells out the name of the *jinja* where the flag was ink stamped and blessed; they say *Minatogawa Jinja.* Directly above that line are two partially faded red ink stamps for the *Minatogawa* Shrine: a square seal below the *kikusui no mon* (half chrysanthemum floating upon river). The third vertical line of characters nearer to the sun's edge are for the name of the flag's recipient. The line of *kanji* reads *Tame* Ishimizu Masaru Kun, or "For Mr. Ishimizu Masaru".

Positioned directly in front of the painted tiger image are two medium sized lines of ideograms that fan out, and away from the sun in typical *yosegaki*-style. Those lines read *Jinchu Hokoku,* or "Seven Lives In Devotion To The Country" and then the name of a relative, Ishimizu Ichinosuke, who probably placed the slogan on the flag as well.

The large tiger head painted upon the upper left-hand side of the flag was done in shades of black, gray, white, red, pink, brown and orange. The artist's brush captured the animal with its focused eyes narrowed, ears back and snarling mouth open.

The tiger head portrait measures 4.5" x 6" (11.43cm x 15.24cm)

The well known print artist Yoshida Toshi (1911-1995), executed the profile of a tiger's head pictured below (c.1926), with noticeable similarities to the image above, captured sometime later by the anonymous good luck flag artist.

Interestingly, when positioned side by side the two tiger portraits – one with mouth open and the other with mouth closed – exemplify the *"O-"* and *"-m"* of the sacred mantra sound *(Om* or *Aum)*, embraced by a number of religions and religious philosophies, including Buddhism.

7. Tigers In Profile

"At the instant a warrior confronts a foe, all things come into focus." – *Ueshiba Morihei*

Tigers in Profile Flag - 27.5" x 32" (69.85cm x 81.28cm)

This unusual silk good luck flag has age darkened and is now a dark cream color. It is marked with a number of prominent holes that were probably caused by mice or rats chewing through the material. It has two thin dark green canvas corner tabs sewn into place, and each of the white silk tie strings remains.

The hoist side edge has been folded over and stitched for added strength. The fly edge has likewise been given added stitching. The red sun in the center was screen printed onto the white field.

The first vertical row of characters on the right-hand side nearest the hoist say, Mr. Sakuradani Iwamatsu; likely the flag's owner. The second line of four characters to the left of that, in conjunction with the single largest character found just below the large square red seal say *Ki Buun Chokyu,* or "I Pray Your Military Fortunes Are Long Lasting". The final vertical line of five smaller *kanji* characters just below those on the right-hand side are a name, Izumi Asajiro. Based upon the location of this name, he may have been the person presenting the flag.

There are two prominent red seals stamped into the right-hand corner of the flag. The square one is large and measures 2.25" (5.72cm) square. Its characters are stylized and difficult to read but may say Sanko Ookami. The second seal is in the shape of a tea leaf. It is 2.25" (5.72cm) long and .75" (1.90cm) across. It appears to read Shiota *Chinju,* or it may say Shiota Village/Town. Shiota Town was located in Saga Prefecture, having been established in 1889. It is well-known for its fields of green tea, and is one of Japan's finest tea producers. The production of green tea in the area dates back to 1504 when it was first introduced from China during the *Ming* dynasty. It recently merged with a number of smaller towns to form Ureshino City.

Near the red sun, just to the left of the 6 o'clock position, a single large character may be observed. This ideogram is pronounced *Riki* or *Chikara,* and means "Power" or "Strength" in English. To the left of this, a similar character is seen. Finally, near the sun in the 11 o'clock position, another large single character, and to its left a smaller one for "Power" or "Strength" is seen. The remainder of the flag contains numerous personal names.

The most prominent feature of this *hinomaru yosegaki* are the two hand-painted tigers that are displayed near the 9 o'clock position, between the 12 and 1 o'clock positions. Rendered in profile, both tiger images are outlined in black ink with white accents as well as red, orange and brown paint. The fact that two tigers were painted on this good luck flag makes it quite colorful and unusual.

Beginning in the Edo period (1603-1868), statues of tigers as well as dragons, wild boar, foxes and other animals were sometimes displayed in pairs, acting as temple or shrine guardians. Functioning much like Chinese protective lions, they are known collectively as *komainu;* their purpose is to keep away evil spirits, and calamity. It is said that the open-mouthed animal is actually forming the first letter of the Sanskrit alphabet *"A",* while the animal with its mouth closed is pronouncing the last letter, *"um".* Together, they say, *"Aum"* which, is probably the most familiar and the greatest of all mantras.

Tiger profile at 9 o'clock position measures 5" x 10" (12.7cm x 25.4cm). The animal's mouth is open in the *"A-"* position

Tiger profile at 12 o'clock position measures 5" x 11" (12.7cm x 27.94cm). The tiger pictured here was painted with its mouth in the closed *"-um"* position. Together they say *"Aum"*.

Tigers in Profile Flag – 28.25" x 40.25" (71.75cm x 102.24cm)

This cream colored good luck flag is made from a silk material that is in superb condition. The hoist side of the flag has two heavy leather tabs sewn into each corner. Each of those continue to have their white silk tie strings in place. The lower right-hand tab is stamped with *kanji* characters that are illegible. The hoist and fly side edges have been hem stitched for added strength.

The red sun or center orb is screen printed onto the white field.

In the 12 o'clock position are two very large horizontal *kanji* characters that say *Hissho*; in English this means "Certain Victory!" To the left of this slogan are two vertical lines of *kanji*. The one immediately to the left of it says *Tame* Hayashi Yukio Kun, or "For Mr. Hayashi Yukio". The vertical line of characters to the left of that read, *Rikugun Taisa,* Kitajima Yonezo (or Yonemitsu). In English these say [Signed By] "Army Colonel, Kitajima Yonezo (or Yonemitsu).

Moving to the approximate 7 o'clock position, nearest the edge of the red sun are two vertical *kanji* characters that again say *Hissho,* or "Certain Victory!" Immediately to the left and slightly above are the vertical characters for "Loyalty and Bravery". To the left of that slogan are another two vertical characters that say *Bu Yu,* or "Valor".

Exhibited along the lower edge of the flag, in the 6 o'clock position, is the profile image of a tiger displayed in front of a bamboo thicket. Both of his front paws have landed upon a rock outcropping, while two unusual circular swirl designs are evident behind him. Bamboo and bamboo grass have symbolic meaning within Japanese culture, and are used in a variety of artistic formats. In some Japanese festivals, these may be used to ward away evil. Overall, bamboo has come to represent virtue, flexibility and strength. Tall and straight, bamboo bends during a windstorm yet will not break. Its upright nature typifies honorableness; its deep root system signifies a resolute character. The interior and exterior of the bamboo symbolizes purity, modesty and innocence. These are admirable qualities within Asian culture, and some of them were codified as necessary traits of the Japanese soldier.

Visible behind and in front of the tiger's head are two design patterns. The example on the left is a clockwise turning spiral or "swirl", while the one on the right depicts a wheel with its center hub and inner/outer rim. Both symbols have their earliest roots in Buddhist spirituality which developed initially in India and then spread throughout Asia. It eventually came to Japan from China, possibly as early as 250 A.D.; its actual practice in Japan is recognized as beginning around 552 A.D.

The swirl of the conch and the wheel are two of the elements that make up what is known as the eight Auspicious Symbols of Buddhism. In early Buddhist tradition, it is said that these eight symbols represent good luck, and were presented as offerings by the great *Vedic* gods to the Buddha upon his attaining enlightenment. Various Buddhist traditions emphasize different meanings, however, their genesis are as follows: The god Brahma (Bonten in Japan), was the first to arrive before the Buddha. He brought with him a golden wheel bearing a thousand spokes. The wheel represents motion, continuity and change; ever moving like the rotating spheres of the heavens. The wheel stands for spiritual transformation brought about through the Buddha's teachings.

Tiger profile with swirl and wheel measures 4" x 7.75" (10.16cm x 19.69cm)

Following *Brahma,* was the god Indra (Taishakuten in Japan), who brought with him a white conch shell horn. In presenting it, he asked the Buddha to loudly announce the truth of the dharma (Buddha's teachings), like a trumpet blast to people everywhere. In similar fashion, India's heroic war gods are often shown holding a rare white clockwise spiraling conch as a symbol of their valorous victories in war. The blast of the conch is said to strike fear into the hearts of one's enemies as well as to drive evil spirits away.

In addition, the right-hand swirl imitates the movement of the sun, planets, and stars about the heavens. Because it has no straight lines, the spiral has come to be a symbol of the continual spiritual growth, and development of a person; a cycle of birth, death, and re-birth. In the earliest depictions, *Brahma,* and *Indra* are portrayed together presenting their gifts of the wheel, and the spiraling conch to the Buddha.

Tigers in Profile Flag - 28″ x 34.5″ (71.25cm x 87.63cm)

This beautiful silk good luck flag has age toned, as silk often does, to a lovely cream color. No corner tabs were ever present on this flag, and the white cotton tie strings are present in both corners. A hem stitch was added across the corner ties in order to re-enforce them to the hoist side of the flag. In addition, extra stitches were placed along the hoist and fly side edges of the flag as well in order to provide more strength to those areas. The dark red colored sun has been screen pressed onto the white field.

Each corner of the flag has one large dark ideogram placed on it. These combine to give the slogan *Buun Chokyu,* or "May Your Military Fortunes Be Long Lasting". Covering the four corners of the flag with this slogan is similar to wishing the owner good battle fortunes in every direction of travel. Note too the highly decorative brush strokes of these characters that replicate the stripes of a tiger's tail.

Across the top of the flag in the 12 o'clock position are a number of vertical lines of *kanji* characters accompanied by five square shrine seals. Some of those lines provide an admonition:

The True Secrets To Victory:

· Let The Enemy Cut Your Cloth, And (You) Cut His Flesh.
· Let The Enemy Cut Your Flesh, And (You) Cut His Bone.
· Let The Enemy Cut Your Bone, And (You) Cut His Body.
· Let The Enemy Cut Your Body, And (You) His Soul/Spirit.

The white field of the flag has 21 shrine seals scattered about it. Each of the stamps are identical and measure approximately 1.25" square. The flag was given to 21 year old Hiroshi Yoshino who hailed from *Shizuoka* Prefecture on the island of Honshu. Some of the lines of characters placed upon this flag are highly stylized special *Shinto* messages, making some of the content difficult to translate into English. Other areas of the flag seem to bare Shinto/Buddhist religious mantras. A mantra is normally recited in order to empty ones mind of distraction. They further act as an aid to bring about a clear and enlightened state of being. Studying the characters that make up a mantra may not actually yield anything with a literal meaning for those normally able to read the ideograms. Only those who recognize the special combination of, or rare pattern of *kanji* will know the actual meaning of the mantra.

Close up photographs illustrate areas where characters have been written on the flag and represent mantras.

The most noticeable character in the center of the image is the standard single *kanji* that makes up one-fourth of the slogan "May Your Military Fortunes Be Long Lasting", or *Buun Chokyu*. Three red shrine seals are visible surrounding it. The upper and the lower right-hand seal have a combination of dark, stylized characters that comprise mantras with hidden meanings.

The lower left-hand aspect of the flag depicts a growling tiger in profile, his mouth opened in a powerful snarl. The tiger's paws are outstretched, while his tail extends up under the lower border of the sun. The tiger profile measures 4.5" x 11" (11.43cm x 27.94cm).

Tigers in Profile Flag – 25.5" x 36.5" (64.77cm x 92.71cm). *Flag courtesy of John Egger*

An interesting tiger-style art flag made from a nice white silk material that is slightly age toned. The reinforcing tabs, placed within the two right-hand corners are made from a heavy, gold foil paper. Each tab has its original white silk tie strings present. As is typical, the hoist and fly side edges of the flag have been slightly folded over and stitched into place for strength. The beautiful red sun in the center of the *hata*, was placed using a silk screen method.

Placed horizontally across the top edge of the flag are five large *kanji* characters that say *Ki Buun Chokyu,* or "I Pray Your Military Fortunes Are Long Lasting". The five medium-large characters written vertically along the left-hand margin of the flag, provide the name of the flag's owner, Mr. Saita Masahira. The rest of the flag is nicely filled with the names of friends and well-wishers of the man who owned the flag.

The full color, highly realistic painted tiger image on this flag takes up more than half of the white field on the right-hand side. With a full paint pallet at the artist's disposal, the tiger was rendered upon the rather large silk cloth patch, using shades of white, black, gray, red, pink, brown, light orange, dark-orange and yellow-gold. When completed, the artist placed his name, two black *kanji* characters, along with his personal seal in red, just below the tiger's tail. The name, and the seal of many poets and artists, are usually tough to read, unless the person is well known. In this case, the characters appear to say *Kyoku-Ko*".

Just above the middle of the tiger's back, two muted *kanji* are visible. The two characters actually form a name that consists of four ideograms, two are partially hidden by the tiger's body. The entire line is visible more fully when the flag is held up against the light. When read, they provide the name, Ota Jinichiro.

Tiger profile patch measures 8.5" x 15" (21.59cm x 38.1cm) with the tiger profile 15.25" long (38.73cm), from the nose to the tail.

8. Tiger Descending

"The fighter is to always be single-minded with one object in view; to fight, looking neither backward nor side wise. To go straight forward in order to crush the enemy is all that is necessary for him." – *Daisetz Suzuki*

Tiger Descending Flag with Blood Oath – 25.5" x 37" (64.77cm x 93.98cm)

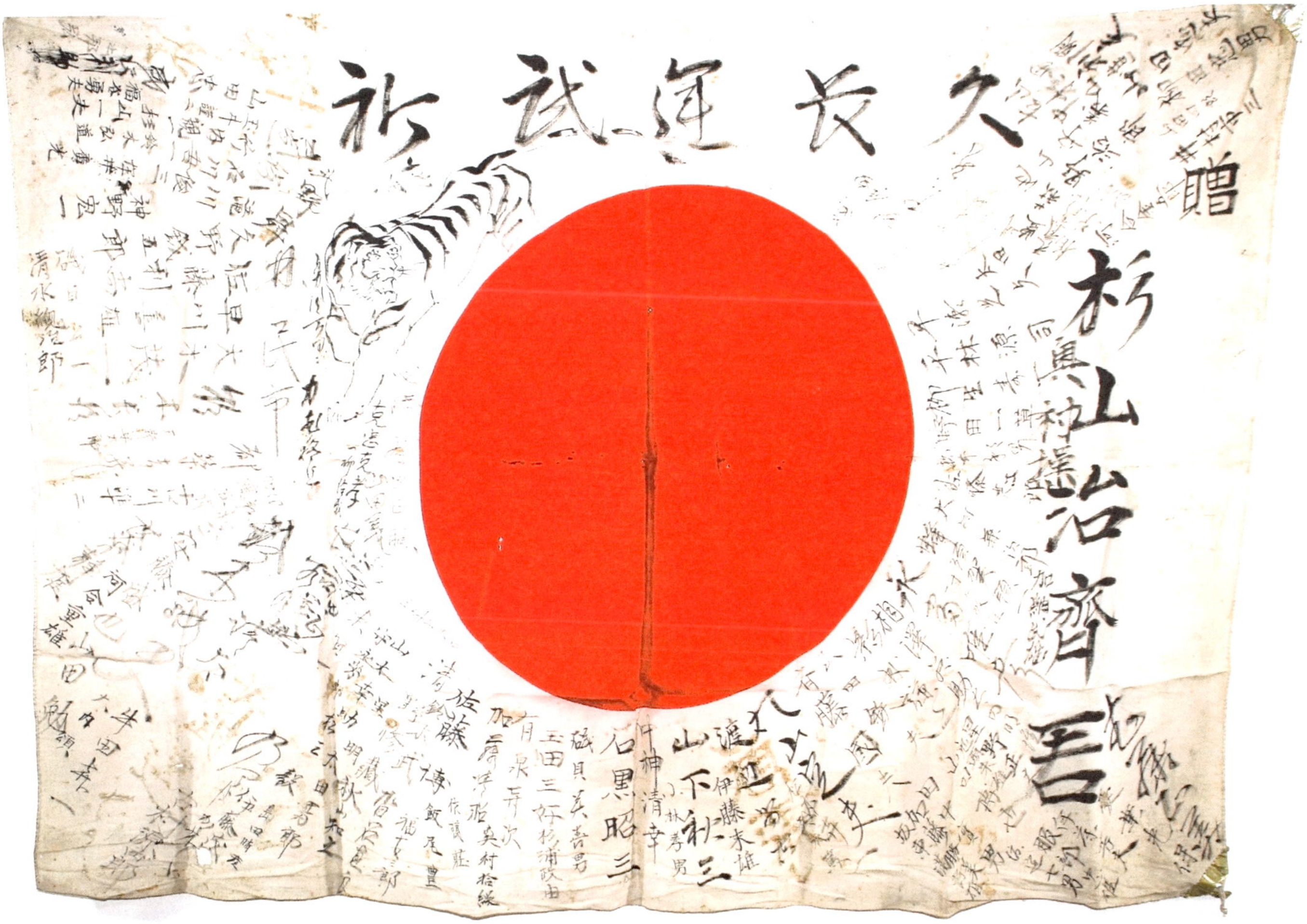

This rare tiger art blood oath flag is made from an age toned silk material. Each of its right-hand corners have been strengthened by sewing triangle shaped gold foil paper tabs into place. Both tabs have their now dirty-white silk tie strings intact. The hoist and fly side of the flag have been provided with additional stitching in order to lend strength to those edges. The bright red center orb has been screen printed to the center of the white field.

Placed horizontally across the top of the flag are five large *kanji* characters that say *Ki Buun Chokyu,* or "I Pray Your Military Fortunes Are Long Lasting". On the right-hand side nearest to the hoist edge are six larger vertical black ideograms (one offset to the right, five directly below), that offer the name of the flag's owner. They say Zo Sugiyama Harunari, or "Presented [to] Mr. Sugiyama Harunari. A smaller but still prominent line of dark characters may be seen to the left of the name of the flag's owner. This is the name of a well-wisher, Okumura Misao.

Near the lower edge of the flag, between the 5-6 o'clock position are visible a

The tiger measures 5.75" x 12" (14.61cm x 30.48cm)

number of vertically placed *kanji* characters that have been written in blood. They are for the slogan *Gambare,* or "Do Your Best". In the upper left-hand corner of the flag near the 11 o'clock position is another blood oath that appears to say *Ki Buun Chokyu,* or "I Pray Your Military Fortunes Are Long Lasting". These ideograms stand out against the normal black inked ones, as they appear rusty brown in color. This blood oath line of characters is one of at least five lines that have been written in like manner across the flag. Similarly, there appears to be at least four round dots or fingerprints placed with blood next to a few of the names that have signed the flag. Finger prints or slogans placed on the flag in blood were intended to imply an even more serious admonishment than what might have been expressed in ink.

Tiger Descending Flag – 28.25" x 31.5" (71.75cm x 80.01cm).

Tiger descending measures 18.5" x 23.75" (46.99cm x 60.32cm)

Made from a thin silk that has toned due to age, this Japanese good luck flag has genuine leather reinforcing tabs in the right-hand corners. Each of the leather tabs has two green silk tie strings fastened into place. The hoist and fly side edges have been strengthened by the addition of stitches. There is no red sun in the center of the flag, although there is a small *hinomaru* above the tiger's rear haunches. The entire design of the flag, except for the black painted *kanji* characters, was applied using a template or block pressed onto the white silk. In the creation of the flag, base colors were applied and allowed to dry, after which successive layers of color were applied over the top, in much the same manner as an *Ukiyo-e* print.

The four red painted *kanji* on either side of the small round sun combine to say *Jinchu Hokoku,* or "Loyalty And Bravery". The long vertical line of ideograms on the fly, or left-hand side, of the flag say *Shuku Nyuei* Minowa Koichi Kun, or "Celebrating Induction For Mr. Koichi Minowa". Positioned down near the lower right-hand corner of the flag are two separate and smaller vertical lines of characters. These say, [Presented In] "Tokyo by Kaneko Tomeyoshi".

On the reverse side of the flag, along the hoist edge is a round red stamp that measures 1.75" (4.44cm), across. The stamp indicates that this item was inspected by a rayon/silk association and says "Pass Approval". This is an extremely colorful "good luck flag", although not a typical *yosegaki*-style version.

The layout and shape of this tiger illustrated good luck signed flag is more in the style of a rectangular send-off or off-to-war banner (*sushei nobori*). While most Japanese send-off banners were long and narrow, like the example shown on page 70, measuring anywhere from 14 inches to 40 feet long, less often seen examples were "flag-shaped", or rectangular.

Rayon/Silk Association "Pass Approval" stamp.

The artwork on the rectangular, or flag-shaped, banner above is superb, and shows a beautiful blue and golden brown kite flying in front of an open lotus blossom. Both the kite and the lotus flower are accented with flecks of colorful glitter, having been custom hand-painted onto the white silk cloth. Just below the bird, the 5 large red horizontal characters read *Ki Buun Chokyu*, or "I Pray Your Military Fortunes Are Long Lasting". This example is even more scarce due to the numerous signatures and messages that were added by friends and well-wishers in the *yosegaki* style. The banner was presented to Abe Hideaki by the Aikoku Life Insurance Company, Ltd., Wada Kaneyasu, Supervisor. That company was merged into the Nippon Life Insurance Company.

The more typical long and narrow off-to-war banner *(Shussei nobori)*, with descending tiger, presented to Mr. Sakauchi Toshio from his friend Miss Hisae. The banner measures: 7.75" x 33.5" (19.68cm x 85.09cm).

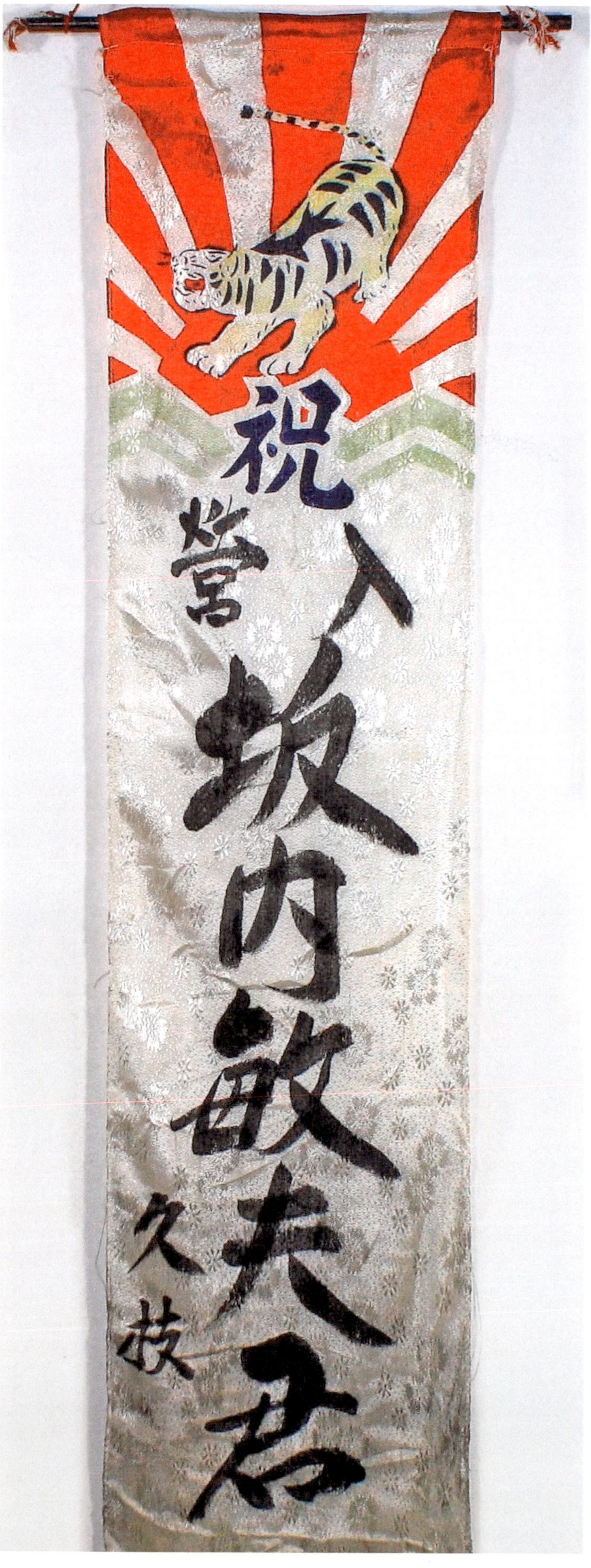

Tiger Descending Flag – 28.25" x 34" (71.76cm x 86.36cm).

This cream colored flag is made from a thin grade of silk. At one time there were two reddish-brown paper corner tabs present, however, only the upper right-hand one remains sewn into place. That same tab also has its white silk tie string remaining. The hoist and fly side edging were reinforced by folding over the rough edges and adding extra stitches for strength. The red-orange sun in the center appears to have been screen printed in the center of the white field.

Tiger descending measures 7.75" x 11" (19.68cm x 27.94cm).

Displayed prominently across the upper portion of the flag are five large *kanji* characters placed horizontally for the war-time slogan *Ki Buun Chokyu,* or "I Pray Your Military Fortunes Are Long Lasting". The upper right-hand area of the flag has two large, vertical rows of heavy, dark characters. The first row on the right gives the name of the flag's owner, Funami Minoru Kun, or "Mr. Funami Minoru". Dropping immediately below that are two medium-large characters that say *Hissho,* or "Certain Victory!" The large line of characters to the left of the owner's name reads *Hissho No Shin Nen* for the slogan, "Faith In Certain Victory!" A smaller vertical line of ideograms just below this says *Go Funto Wo Inoru,* or "Praying For A Strong Fight."

Near the edge of the center orb, between the four and five o'clock position and coming out in typical *yosegaki* "sideways" fashion, are three rather dark, medium-sized characters. Just to the left of those are four additional characters that say *Yusen Funto* or, "Fighting The Battle With Courage". A common slogan often seen lies along the sun's edge, in a vertical position between the nine and ten o'clock position. The four large, heavy characters say *Shisei Hoko,* or "Serving With Utmost Sincerity."

Tiger Descending Flag – 28″ x 34.25″ (71.12cm x 86.99cm).

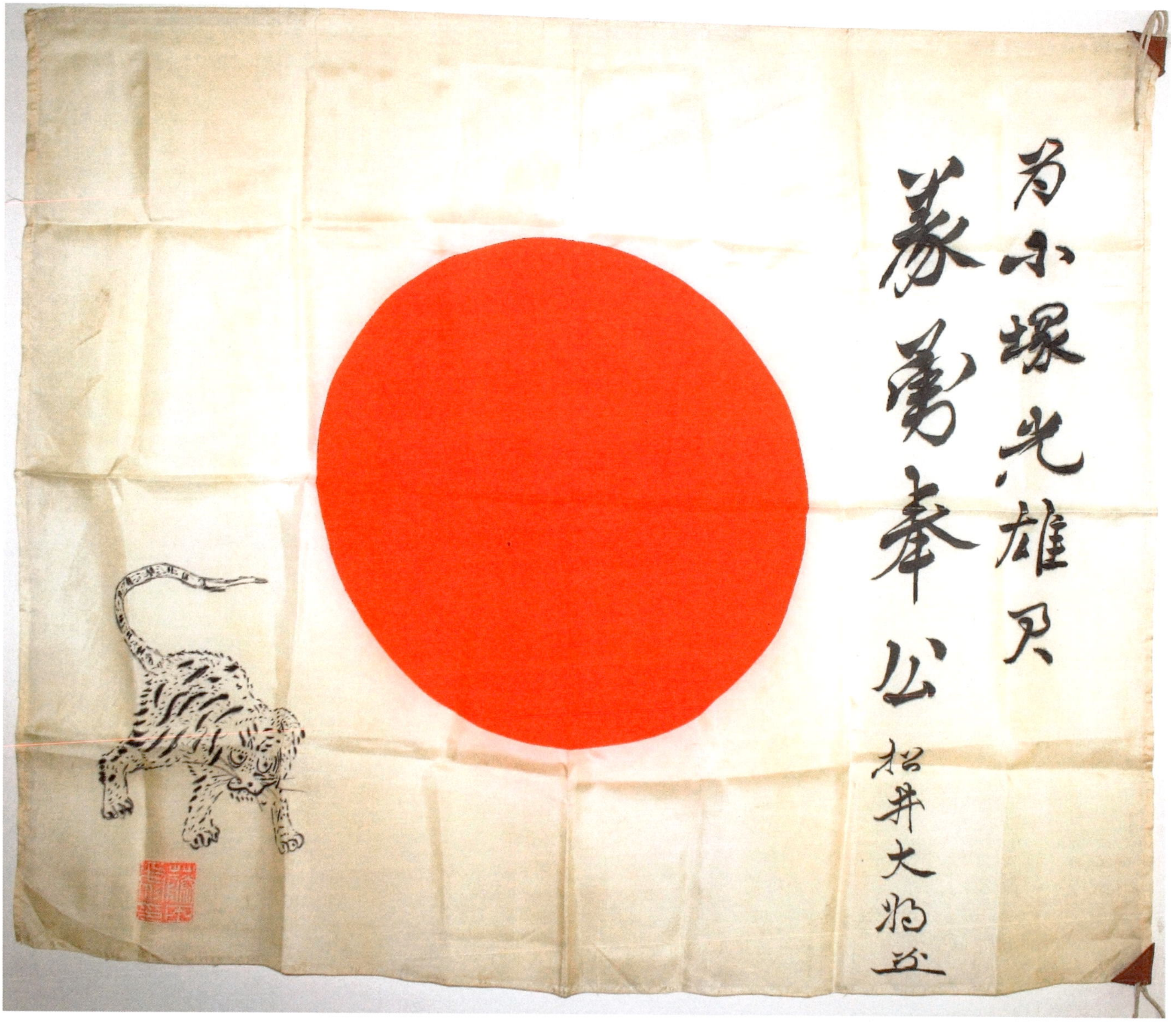

This thin silk good luck flag has darkened with age to a light tan color. The two right-hand corners each have a thin leather reinforcing tab sewn into place, and each of those has their white silk tie strings securely fastened. The fly and hoist side edges have accessory stitches added to them for strength. The red sun in the center of the white field has been screen pressed into place.

Tiger descending measures 7.75" x 9" (19.68cm x 22.86cm).

The vertical line of six *kanji* characters written nearest to the hoist side edge say *Tame* Kozuka Mitsuo Kun, or "For Mr. Mitsuo Kozuka. The bold vertical line of ideograms to the left of that say *Giyu Hoko,* or "Patriotic Courage And Loyalty". The smaller vertical line of five *kanji* immediately below that say, Matsui Taro, the name of the person signing the above slogan. The red seal below the tiger measures approximately 1.75" (4.44cm) square and is illegible.

The black and white "kitty kat" style tiger image painted on the good luck flag above is reminiscent of the animal created here by the *Ukiyo-e* artist Yoshikazu, who was active c.1850-1870. Note the similarities between the head, body and tail positions

Below: Colorful *Ukiyo-e* print of a tiger by the artist Ichijusai Yoshikazu, c.1848-1870

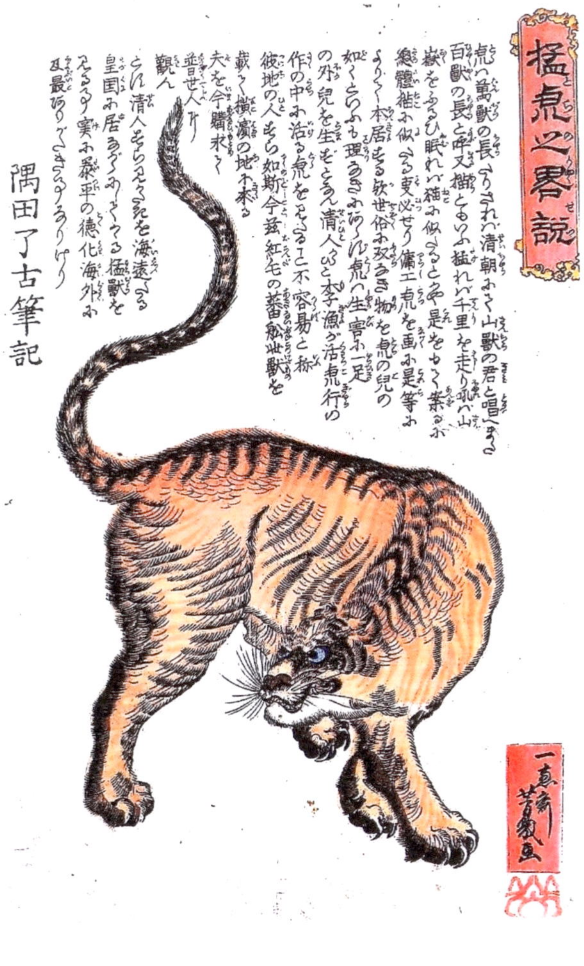

Tiger Descending Flag – 27" x 31" (68.58cm x 78.74cm).

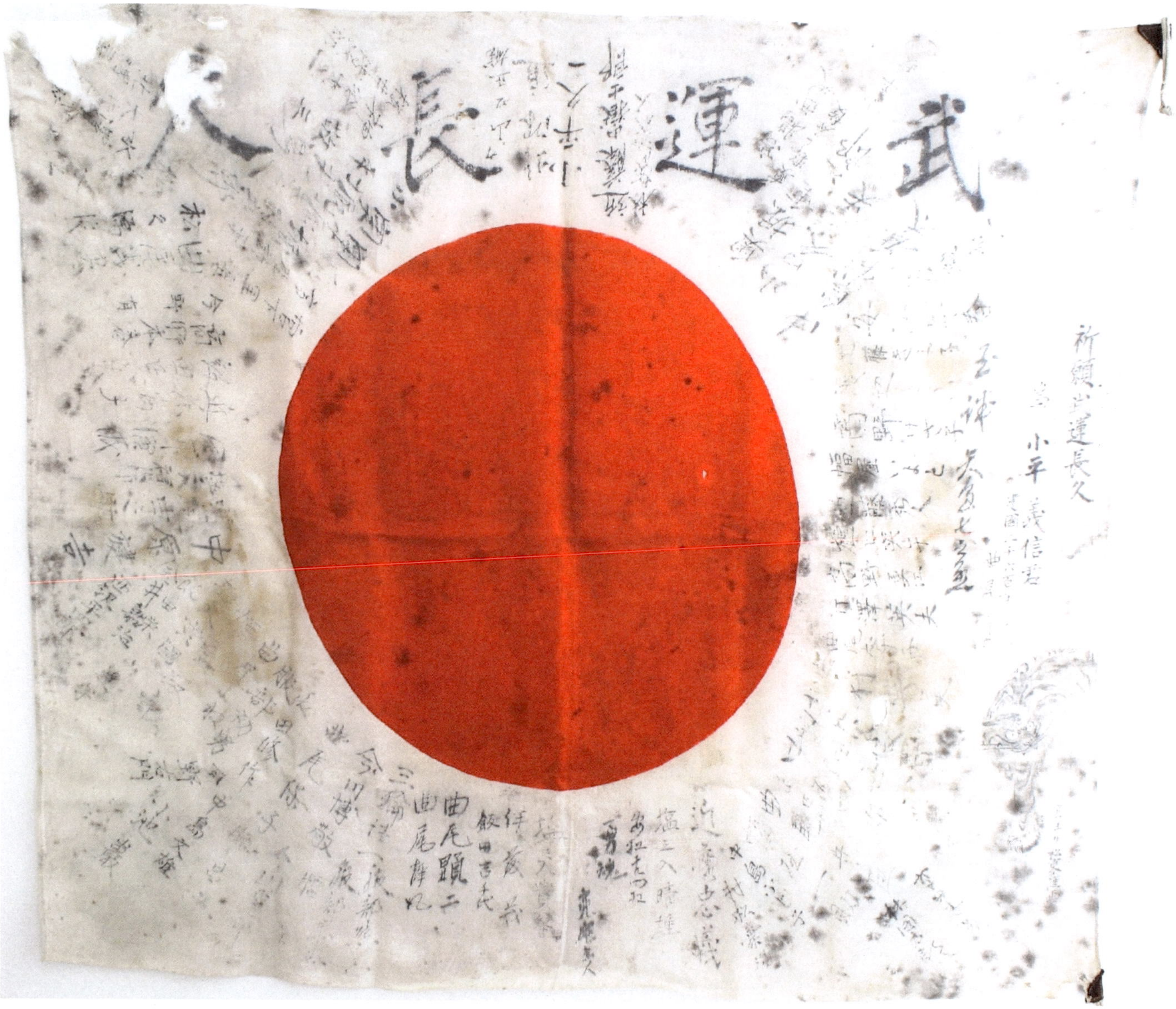

This white silk Japanese good luck signed flag shows various tears and stains associated with the affects from use in the field, as well as post-capture cleaning and later storage. The hoist side of the flag maintains both leather corner tabs and white silk tie strings. The hoist and side edges of the flag have been hem reinforced for added strength. The red sun was screen printed near the center of the white field.

The flag is covered with *kanji* characters, although many of them are faded and difficult to read. Quite visible across the top aspect, however, are the four large horizontal ideograms that read *Buun Chokyu,* or "May Your Military Fortunes Be Long Lasting". Five signatures may be seen inked near the 12 o'clock position and have been placed upside down. The second one from the right is the name Kondo Takeshiro.

Moving clockwise, and closest to the the hoist side of the flag, are six prominent vertical lines of *kanji*. Those may be seen to surround the black and white image of an old-style tiger that appears to be descending. The outermost line of characters contains six *kanji* that are similar to the more standard five character slogan for *Ki Buun Chokyu,* or "I Pray Your Military Fortunes Are Long Lasting". In fact, this "new" slogan says *Ki-gan Bu-un Cho-kyu* and is just a longer way to say "Fervent Prayers For Everlasting Battle Fortunes". Directly below that, positioned to the right of the tiger, is the signature of the artist that painted the beast. The ideograms read, "Drawn With Respect By Shoto" (or Matsuhigashi, depending upon how the artist actually pronounced his own name), "Born In The Year Of The Tiger".

The next line of characters, directly to the left and above the tiger, gives the name of the flag's owner, "For Mr. Kodaira

Yoshinobu". Immediately to the left of the recipient's name is the presentation date, "1st Day Of Year 2600 (1940). The number 2600 follows the old *koki*, the Japanese imperial or national calendar year designation. Interestingly, the 1940 Summer Olympics and Tokyo Expo were planned as anniversary events this same year. The famous Japanese "Zero" fighter also took its "0" designation from the last number in *koki* 260(0). Written vertically to the left of the date line is a name, Magario Moritsugu, and to the left of that another name, Yajima Hichinosuke.

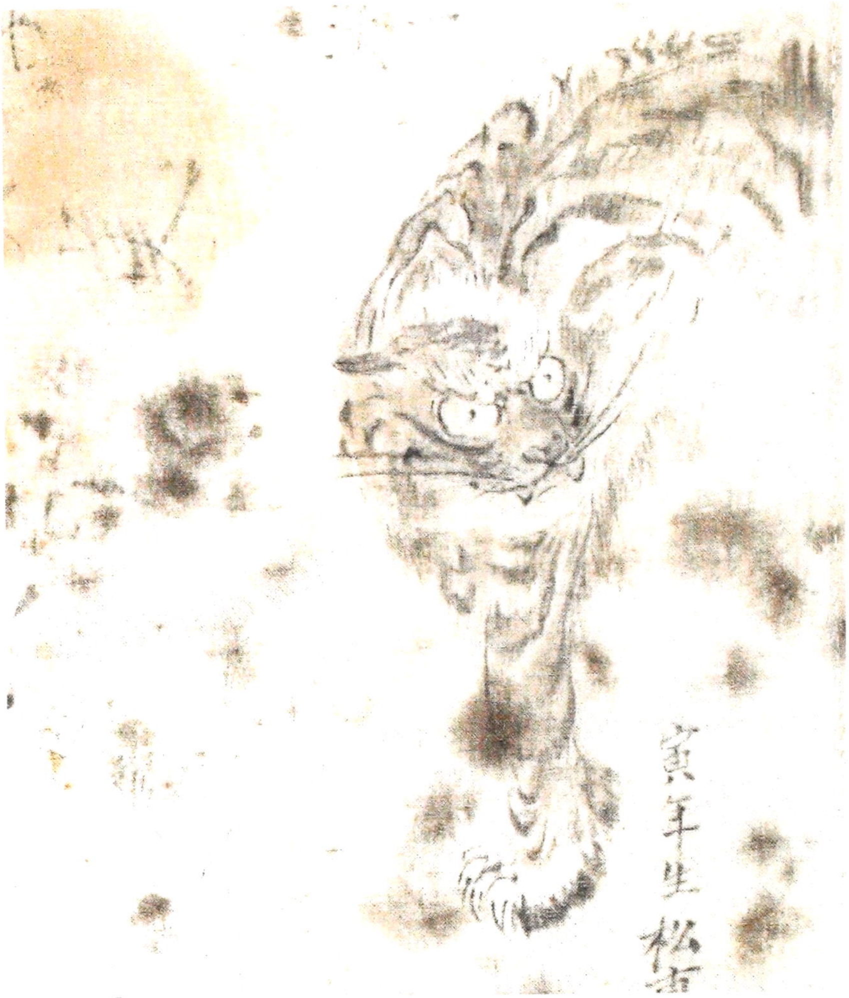

Detail of tiger descending measures 6" x 3.5" (15.24cm x 8.89cm).

"Tiger in Bamboo Forrest", c.1810 *Ukiyo-e* print by Kikugawa Eizan, (1787-1867)

This tiger was painted in an old fashioned style, meant to convey the animal's movement; paws crossed, he appears to be in motion. Other elements are common to this particular genre as well: for example, the beast has large, round eyes that depict his focus and intent. In addition, his coat is striped in similar fashion to others of the period. In comparison, the same elements are nearly identical in the highly sought after, "Tiger in Bamboo Forrest" *Ukiyo-e* image done by the famous artist, Eizan.

The flag shown above was a "Bring Back" souvenir by an American veteran who served overseas in Luzon, Philippine Islands. His story follows: Inducted into the United States Army on December 16, 1942 he was assigned to various companies before being transferred from Fort Lewis, Washington to Camp Maxey near Paris, Texas on or around July 1, 1944. In describing its military complex, the Camp Maxey Public Relations Office states that the facilities are, "...ideal for working out Infantry problems...and training for modern battle conditions." Once there, he was assigned to the 1267th Combat Engineers before shipping out.

A view of Fort Lewis, Washington.

Reminiscing about his dad, the soldier's son said that after the war, his father often spoke about Japanese General Yamashita Tomoyuki, the Tiger of Malaya, whenever he would take the flag out and discuss it. In a letter written home to his wife in October 1945 from Luzon, Philippine Islands, one month after the Japanese surrender, the soldier identifies this exact flag, and describes how he acquired it from a local *Igorot* tribesman. The name *Igorot* describes collectively a group of Austronesian peoples that inhabit the mountainous regions of Luzon, Philippine Islands. Please note: the language of the time used to describe the Japanese troops may be offensive. In historical review, the slang conveys the common thinking of the World War Two era, where enemies on both sides were completely dehumanized in order to focus attention on total victory.

Soldiers line up in ranks at Camp Maxey, Texas.

In a letter home, a U.S. serviceman describes how he acquired his tiger painted flag. In part, the letter reads: "The flag I got from an *Igorot* who had taken it from a Jap. Almost all the Japs carried flags, and they valued them highly. This one is a little battle worn, as they all are. I washed and boiled it in G.I. soap, but not all the stains came out, its clean tho'. Yamishito (sp.) was known as the "tiger of Malaya" and if you hold this flag up by the leather tips and look along the top edge you will discover a tiger."

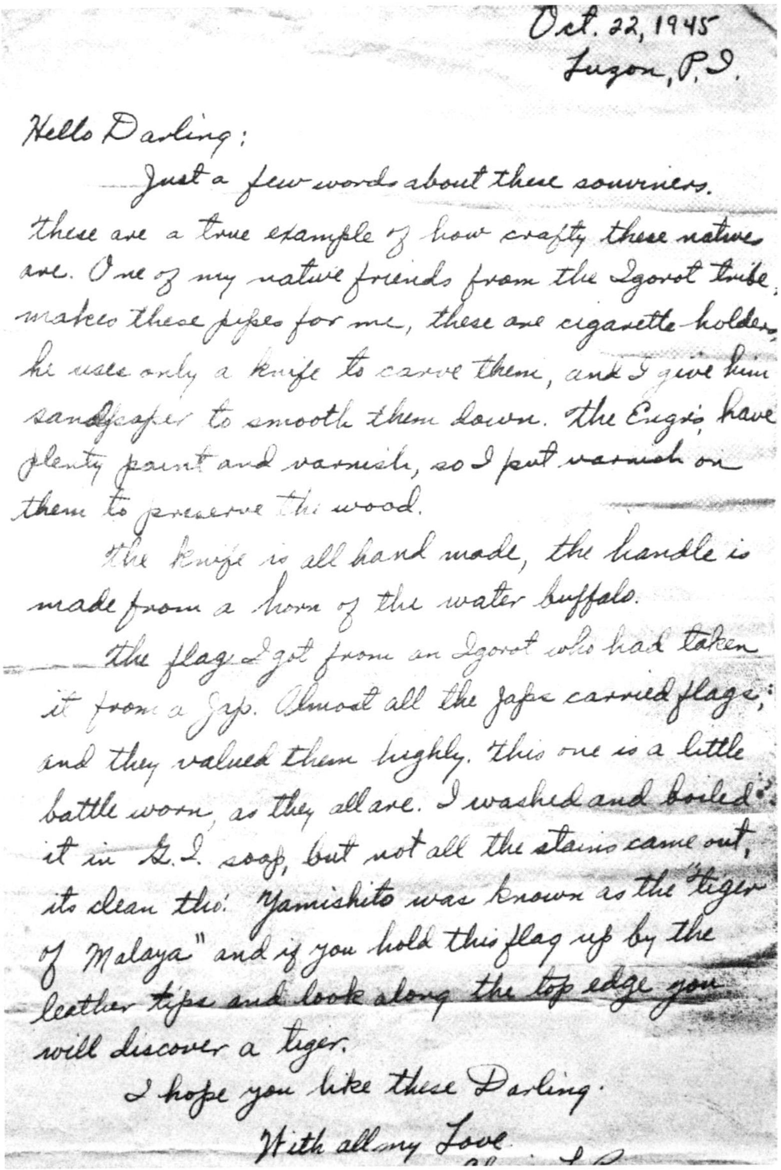

Oct. 22, 1945
Luzon, P.I.

Hello Darling:

Just a few words about these souvenirs. these are a true example of how crafty these natives are. One of my native friends from the Igorot tribe makes these pipes for me, these are cigarette holders he uses only a knife to carve them, and I give him sandpaper to smooth them down. the Engrs. have plenty paint and varnish, so I put varnish on them to preserve the wood.

the knife is all hand made, the handle is made from a horn of the water buffalo.

the flag I got from an Igorot who had taken it from a Jap. Almost all the Japs carried flags; and they valued them highly. this one is a little battle worn, as they all are. I washed and boiled it in G.I. soap, but not all the stains came out, its clean tho'. Yamishito was known as the "tiger of Malaya" and if you hold this flag up by the leather tips and look along the top edge you will discover a tiger.

I hope you like these Darling.

With all my Love

The caption on this image reads, "An *Igorot* soldier and his wife."

Yamashita Tomoyuki was placed in charge of the Japanese 25th Army on November 6, 1941. On December 8 his troops invaded Malaya, and in 70 days speedily concluded the campaign with the conquest of Singapore on February 15, 1942. British troops were so soundly defeated that Prime Minister Winston Churchill remarked that the loss was the "worst disaster" in British military history. Yamashita's stunning success earned him the title, "The Tiger of Malaya".

Near War's end, his superiors ordered General Yamashita to the Philippines in order to defend them against advancing American and Filipino forces. Japanese troops gave ground begrudgingly, and while their overall victory was doomed, large portions of Luzon remained in Japanese hands until after the formal surrender took place aboard the U.S.S. Missouri in Tokyo Bay on September 2, 1945.

Japanese General Yamashita Tomoyuki, "The Tiger of Malaya". The General's sword is in the collection of the United States Military Academy Museum at West Point, NY.

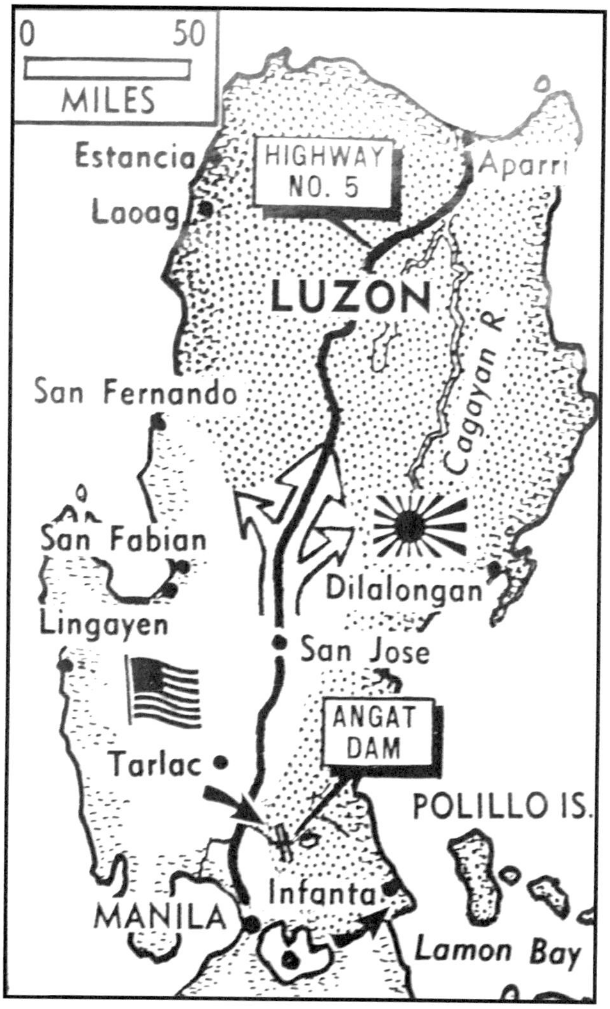

From an Acme Photo dated 5/16/45, "Fighting Still Goes On In Luzon: American forces are driving up highway No. 5 from San Jose against General Yamashita's Japs defending Northern Luzon and the Cagayan Valley."

9. Tiger Standing On Rock Ledges

"So long as there is breath in me, that long will I persist. For now I know one of the greatest principles of success; If I persist long enough I will win." – *Og Mandino*

Rock outcroppings and boulders are often prominently matched along with the tiger and bamboo in Asian art. These contain special symbology, and while they may represent a number of things, rocks in the Zen garden are especially meant to deliver a message. For example, large rocks may stand for mountains, while small ones protruding above a base of even smaller raked pebbles may be emblematic of baby tigers.

During the *Heian* Period (784-1185), rocks were often used as elements that represented political situations and/or commentary. Japan's oldest book on gardening, the *Sakuteiki,* or "Secret Extracts on Gardens", provides this interesting account: "Sometimes, when mountains are weak, they are without fail destroyed by water. It is, in other words, as if subjects had attacked their emperor. A mountain (the emperor), is weak, if it does not have stones (for example, wise advisors and a sympathetic populace), for support." [It is because of...] "stones that a mountain is sure, and thanks to his subjects that an emperor is secure. It is for this reason that, when you construct a landscape, you must at all costs place rocks around the mountain."

Based upon this story, the analogy is quite clear. In the same manner that the wise counselor and devoted military servant play pivotal support roles with specific qualities that maintain a nation's leadership, similarly the strong and watchful eye of the tiger is seen as vital to the security of the mountain against any storms (i.e. difficult situations or enemies), that might threaten to sweep it away.

Tiger Standing on Rocks Flag - 29" x 31.25" (73.66cm x 79.37cm).

This good luck flag is made from thin silk and retains most of its original bright white color and sheen. While in good condition, it has a number of tears in the material, primarily in the 12 and 3 o'clock positions. There are two genuine soft leather corner tabs sewn into place, however, only the lower left-hand silk tie strings remain. As in most good luck flags, the hoist and fly side edges have been reinforced with additional stitches. The red sun was likely added via a silk screen process.

There are three lines of large, dark vertical *kanji* on the right-hand side of the flag. The line closest to the hoist side says *Hichisei Metteki,* or "Seven Lives To Decimate The Enemy". The second line just to the left of that says *Kaigun Chujo* Sugiyama Toshisuke, or "Navy Vice-Admiral Sugiyama Toshisuke". The third and longest line of ideograms, closest to the red *hinomaru,* dedicate the flag to its owner; they say *Shuku Shussei* Hirose Chokichi Kun or "Celebration For Going Off To Battle Mr. Hirose Chokichi". On the opposite side of the flag, near the fly side edge, is a line of five *kanji* characters and two red seals that pass through the tiger's tail. The ideograms provide the name of the artist, Yamaguchi Yoshikazu *Ga*. In English they say, "Drawn By Yamaguchi Yoshikazu".

Each red seal measures 0.75" x 0.75" (1.9cm x 1.9cm). The upper seal is the name of the artist, Yamaguchi Yoshikazu. The lower seal appears to be the pen name of the artist, Ryusei.

Tiger standing on rocks measures 14.5" x 8.50" (36.83cm x 21.59cm).

Immediately below the lower red seal are two lines of *kanji* characters: a thinner line of five ideograms on the left, and to its right, a group of three bold *kanji.* These combine to say *Nihonto,* or "Japanese Sword." The Japanese *nihonto* was the samurai warrior's weapon of choice, and was said to embody the essence of the warrior's spirit. That same quality was undoubtedly felt by those stationed aboard not only the *Musashi,* but other ships as well. In general, the weapon was the embodiment of the warrior, and would become a key to his ultimate victory over the Allies. Just to the right of that, five thin characters may be seen grouped together to form the slogan *Ki Buun Chokyu,* or "I Pray Your Military Fortunes Are Long Lasting". The realistic tiger standing upon a rocky outcropping is painted in varying shades of black, gray, white, red and red-brown. The shading, especially upon his head, upper body, front legs and body show great detail, really bringing the image to life.

Japanese Vice-Admiral Sugiyama Toshisuke *(1886-1946),* was an interesting and important figure both before and during World War Two. He was a graduate of the Japanese Naval Academy at Etajima, Class 35 on November 20, 1907. In the years that led up to the Great East Asia War, he rose through the ranks in various positions, eventually becoming Commander of the Navy Technical Department. He was a proponent of large capital ships and was instrumental in making certain that funds were available to build the Japanese super battleships, *Yamato* and *Musashi.* His often behind the scenes maneuvering limited the efforts of his strategic competition within the military, who believed that funding these huge battleships would create "white elephants" and be a waste of money.

Vice-Admiral Sugiyama Toshisuke, seated front row on board the super battleship *Musashi.* The photograph was taken during an Imperial visit, or *gyoko,* on June 24, 1943. Although the Emperor's visit was supposed to be "TOP SECRET", his imperial flag was nevertheless hoisted over the ship, contravening official regulations for all to see. The ship was based at the Yokosuka Naval Base. Hirohito is seated in the front row, center.

Tiger Standing on Rocks Flag – 26" x 40.5" (66cm x 102.87cm). *Flag Courtesy of John Egger*

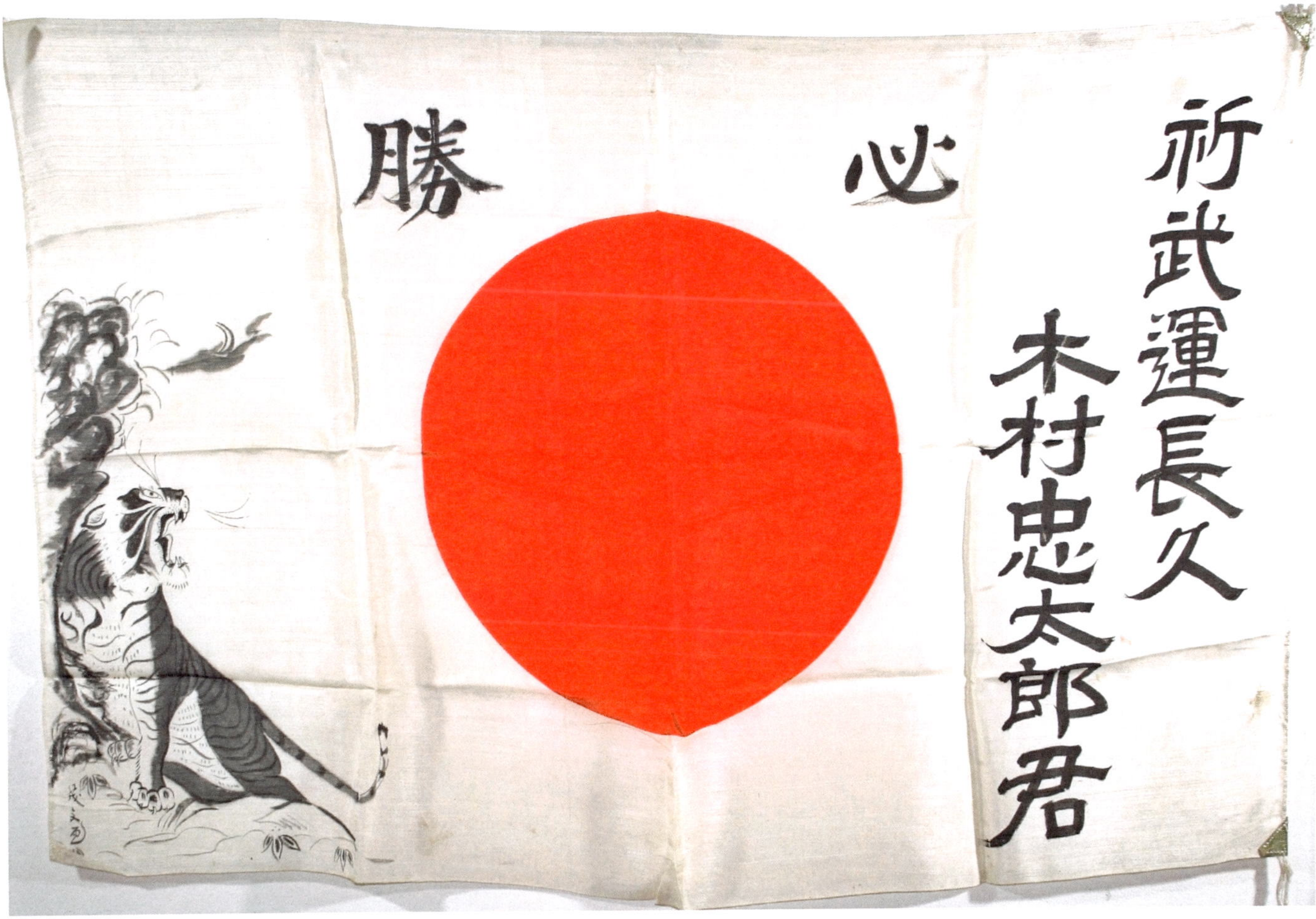

Made from a thin, stiff silk material, this good luck flag has two gold foil paper corner tabs sewn into place on the right-hand side. Wear and age have caused the tabs to lose much of their gold color, and they now appear to be a shade of silver. Both tabs also have their white silk tie strings securely fastened in place. The hoist and fly side of the flag are both stitched with additional thread for strength, and the red center orb has been screened onto the white field. The large tiger, standing upon a rocky slope in the lower left-hand corner, was painted using shades of black and gray ink.

Two bold characters may be seen drawn across the top or 12 o'clock position of the flag. These represent the slogan *Hissho,* or "Certain Victory!" There is a definite implication that accompanies this slogan, and that involves hard work in order to achieve the goal. On the right-hand side of the flag, closest to the hoist side edge, are five well written characters for the slogan *Ki Buun*

Tiger standing on rocks measures 17" x 11" (43.18cm x 27.94cm).

Chokyu, or "I Pray Your Military Fortune Are Long Lasting". It is probable that the same person wrote all of the *kanji* on this flag; their writing style was neat, bold and impressive. To the left of that slogan, the name of the owner of the flag is given, "To Mr. Kimura Chutaro".

Tiger Standing on Rocks Flag – 28.5" x 38.25" (72cm x 97cm). *Flag Courtesy of John Egger*

This magnificent full color tiger art flag is made from a heavy silk material. It has unusual corner reinforcing tabs – triangular shaped pieces of fancy embroidered cloth have been sewn over the existing white silk material of the flag. The colorful material is reminiscent of that sometimes seen on the robes of Buddhist monks or *Shinto* priests. In addition, white silk tie strings are found in both right-hand corners and are securely stitched in place. The fly and hoist side edges (in this case left and right-hand sides respectively), have been folded under and double rows of stitches added for strength. The red *hinomaru,* or sun, was probably screen printed onto the white field.

Boldly brushed across the top of the flag in five horizontal ideograms is the slogan, "I Pray Your Military Fortunes Are Long Lasting", or *Ki Buun Chokyu.* In the upper left-hand corner, adjacent to the aforementioned slogan, is a vertical line of six medium sized *kanji* characters. This provides the dedication of the flag and reads "Presented To Mr. Takahata Yoshinobu". The rest of the flag contains virtually nothing but names.

Detail of flag opposite. The tiger standing on rocks measures 14" x 14.5" (35.56cm x 36.83cm).

Tiger Standing on Rocks Flag - 25.25" x 35.5" (64.13cm x 90.17cm).

Pictured is a white silk good luck flag that is darkened due to normal age and smearing from the black ink signatures. The right-hand corners have brown paper tabs sewn into place; these were made to simulate brown leather. Both tabs have their silk tie strings securely stitched into place; the examples shown here are an unusual black color. The bright red sun in the center of the field was silk screened onto the background.

Ringing the entire outer rim of the red center orb are the single *kanji* characters for the word *riki,* or *chikara,* which means "power" in Japanese. This implies that the flag's owner should be strong and firm in doing his duty while away. In addition, it indicates a compounding of that wish for added strength from those that have placed that *kanji* character upon the flag. Along the hoist side edge of the flag are five prominent vertical characters that say *Ki Buun Chokyu,* or "I Pray Your Military Fortunes Are Long Lasting". To the left of that is another vertical line of *kanji* that present the flag to its owner. The characters say Fukushima Katsuichi-Kun, or Mr. Katsuichi Fukushima.

Positioned just to the left of the previous line of characters, and slightly lower is a grouping of four smaller *kanji* characters that say *Zo Dai-Ni-Ku* or "Presented By Zone #2". In the lower section of the flag, observed between the 4-5 o'clock positions, are multiple signatures from members of the Kobayashi clan.

Of additional interest are 13 separate repairs to the silk material that were likely performed in the field, or done after returning home. Similar to others observed in the past by the author, these were accomplished using a khaki-colored, heavy cotton thread. One of the repairs may be seen in the close-up image of the tiger on page 87, located across his hind quarters and tail.

The black and white tiger painted upon this flag is somewhat of a transition in terms of artistic style. The body aspect appears to have been fashioned more similarly to those painted by the Japanese Zen monk, Sesson Shukei (1504-1589).

While primarily a landscape artist, Sesson did leave behind a body of work that portrayed animals of various kinds including birds and tigers. The animal's head, on the other hand, contains a bit more realism; the kind we might expect to see in modern tiger art.

Right: Tiger standing on rocks portrayed on flag opposite measures 5.5" x 5.5" (13.97cm x 13.97cm).

Correspondingly, the *Ukiyo-e* artist Kitagawa Utamaro I, captured comparable aspects of this fanciful, old tiger painting style: the tiger's crossed paws, hunched shoulders, curved striped coat, cocked head and growling (almost smiling) features that were intended to give the animal muscular tension and movement.

"Tiger" *Ukiyo-e* print, c.1800 by Kitagawa Utamaro I, (1753-1806).

Tiger Standing on Rocks Flag - 27.75" x 31" (70.48cm x 78.74cm).

This image shows a heavily worn Japanese silk good luck signed flag with some interesting content. The khaki-brown cloth corner tabs remain firmly sewn in place on the right-hand, or hoist, side and each has its original silk tie strings affixed. Both the hoist and fly side edges have supplemental stitching in place for extra strength. The red screen-printed sun in the center of the white field retains its original deep color. There are two noticeable rips on the left-hand aspect of the flag that have been mended back together: the lower one is approximately 12" long (30.48cm), while the upper one is about 10" long (25.4cm). Both repairs were made using red thread that was loosely zigzag stitched in order to rejoin the two sides.

Centered horizontally across the top of the flag are five moderately sized *kanji* characters that say *Ki Buun Chokyu,* or "I Pray Your Military Fortunes Are Long Lasting". Immediately below those, near the 1 o'clock position are two larger characters for the slogan *Hissho,* or "Certain Victory!" To the right-hand side of the previous phrase, nearest to the flag's hoist edge, is a vertical line of six large ideograms that name the man that owned the flag; they say, "Presented to Matsumoto Shouji". To the left of the presentation line is a smaller vertical line of *kanji* between the 3-4 o'clock positions. Those characters read *Ki Buun Chokyu,* or "I Pray Your Military Fortunes Are Long Lasting".

The vertical line of characters and small rendering of a *sake* cup, located in the 6 o'clock position, illustrate a typical Japanese pun, or "play on words". The text reads *Genki Ippai,* which is a common phrase meaning "Filled With Happiness", or "Brimming With Spunk". *Ippai,* however, may also mean "One Cup", so the replacement of it with the image of a *sake* cup is a double-entendre signifying "Filled With Sake". Counting two vertical lines of *kanji* characters to the right of the sake cup, reveal four dark, vertical ideograms that say "Serving With Justice And Bravery". Moving clockwise again, the single large *kanji* character just beyond the 6 o'clock position near the sun's red border says "Fidelity".

In the flag's lower left-hand corner is the black, white, orange and red painted image of a tiger, emerging from its bamboo hiding place and lunging toward the edge of a rocky precipice. The tiger's ears are angrily pulled back, its mouth frozen open in a snarl. Below the tiger are three vertical lines of characters. Translated they say "Travel 1000 Miles To Vanquish The Enemy, Then Travel 1000 Miles Back [Home]". This saying alludes to the popular Japanese sentiment that a tiger had the ability to travel a thousand miles, out and back in search of prey, always returning safely home. Moving counter-clockwise from the tiger, just to the right of the bamboo are two small vertical *kanji* characters that probably identify the artist by his last name, Anzai.

Between the 9-12 o'clock positions are numerous names and slogans. The two dark, medium-size characters just above the 9 o'clock position say *Chuyu,* or "Loyalty And Bravery". Moving clockwise near the 10 o'clock position are the four medium-size *kanji* that read *Buun Chokyu,* or "May Your Military Fortunes Be Long Lasting".

Tiger standing on rocks measures 5.5" x 8" (13.97cm x 20.32cm).

Tiger Standing on Rocks Flag – 27.25″ x 38″ (69.21cm x 96.52cm).

This silk Japanese good luck signed flag catches the eye due to its fine condition as well as its strikingly colorful tiger illustration. The brown paper tabs, made to look like leather, are sewn into each right-hand corner. In addtion, the heavy silk tie cords remain firmly knotted in place. The hoist and fly side edges of silk were folded under, while stitches were applied to create a hem for strength. The red center sun is a deep shade of red, and is silk screened onto the white background.

The front of the flag contains over 100 signatures of friends and well-wishers. In fact, nearly all of the *kanji* characters are for personal names. There are, however, a few interesting patriotic Japanese slogans present. Placed boldly across the 12 o'clock aspect of the flag are four prominent *kanji* ideograms. The slogan says "Destroy America And Britain!" Those characters may be difficult to see initially because two or three personal signatures have been added, and intersect them. In front of the tiger, five very bold *kanji* characters are placed vertically, and they say "I Pray Your Military Fortunes Are Long Lasting," or *Ki Buun Chokyu*. Just below and to the right of that, in the 3 o'clock position, is another slogan composed of four bold *kanji*; those characters say "Serving With Justice And Bravery".

Across from the previous slogan, also in the 3 o'clock section of the white field, are three more lines of dark, vertical ideograms. The center line contains five characters, and once more says *Ki Buun Chokyu,* or "I Pray Your Military Fortunes Are Long Lasting". To the left and right of that are personal names, Furuto Kumetaro and Imahashi Junzo.

Located in the 6 o'clock position along the lower edge of the flag, someone has signed their name in a flamboyant style of script. The name appears to be for a man and says Takamatsu Ichizo. Continuing clockwise to the lower left-hand corner, a friend has added two vertical lines of characters for the phrase "Pursue The Utmost Sincerity With Life" [Signed] Asaina Yasuaki.

The marvelous illustration of a tiger, standing upon rocks amid a bamboo grove, is painted in full color. For his life-like depiction, the artist chose a full palette of shades including: black, white, red, orange-brown, yellow, gold, green and blue ink/

paint. Below the tiger's paw, along the flag's hoist side edge, the artist signed his name, Hiromasa, and placed his personal red ink stamp below it. The stamp measures 0.5" square (1.27cm). Crowning the painting, the artist provided a final comment, "Born In The Year Of The Tiger". The tiger is the third in the twelve year cycle of Chinese zodiac signs. Leading up to World War Two, the following years were celebrated as tiger years: 1914, 1926 and 1938.

Tiger standing on rocks measures 4" x 7" (10.16cm x 17.78cm).

10. Tigers On The Attack

"One finds life through conquering the fear of death within one's mind. Empty the mind of all forms of attachment, to make a go-for-broke charge and conquer the opponent with one decisive slash." - *Togo Shigekata*

Tiger Attacking Flag – 40" x 42.25" (101.6cm x 107.32cm). *Flag Courtesy of Dean Kanabrocki*

This magnificent and quite large white cotton presentation flag was a handmade gift to a Japanese officer. The flag has two large pieces of white triangular shaped cotton cloth sewn to the reverse, in the right and left-hand corners for added strength. Each of those tabs has white cotton tie strings in place. The hoist and fly side edges have been heavily stitched for increased strength. The large center sun was outlined in pencil and then colored in with red paint.

The three most prominent *kanji* characters seen applied horizontally across the top of the flag say *Ki Zensho,* which means "Prayers For Complete Victory!" Immediately to the right of that, the vertical line of six *kanji* characters says "For Commander Oshima", the man receiving the flag. Another single line of vertical ideograms lies just below the tiger's tail reads *Issei Banteki Iatsu,* or "Dominate The Enemies With One Shout!".

On the lower right-hand corner of the flag, appearing almost chiseled into the rock base are several lines of *kanji* that provide a terrific dedication: "This Work Is Presented To Show The Most Sincere And Deepest Gratitude, Begging Your Kind Acceptance. In Commemoration Of The 'Day of Meiji' (Birthday Of Emperor Meiji), *Showa* 12 (1937), By Shundo Hachio".

Tiger attacking measures 37.5" x 32.5" (95.25cm x 82.55cm).

Tiger Attacking Flag – 27" x 36" (68.58cm x 91.44cm).

Often good luck flags made from silk will be found to have darkened with time. This particular flag is no longer white, yet has a number of beautiful qualities that continue to stand out when viewed. Its two brown paper corner tabs, made to appear as leather, are well sewn into place. There are two colorful gold silk tie strings, rather than the typical white ones, attached to each of the tabs. Interestingly, this flag never had hem stitched hoist and fly side edges. Instead, its upper and lower side edges have been provided with extra stitched lines. The deep red colored orb in the center of the white field is silk-screened.

The flag's two largest characters, seen in the 12 o'clock position, say *Hissho*, or "Certain Victory!" Along the right-hand side of the flag is a vertical line of medium sized characters. The seven *kanji* give the name of the flag's owner, "For Mr. Yamamoto Kimizo".

Positioned prominently across the top of the flag, near the edge of the red sun, are three shrine or temple seals. These measure approximately 2.5" (6.35cm) square. The ink used to make the seals spread out after application, and the characters identifying the place name are difficult now to read. Buddhist and *Shinto* places of worship normally have a number of different locations within the complex where personal or organized worship may take place. Often there is an outer, middle and inner sanctuary where different ceremonies or practices occur. It is likely that these seals are representative of such places. The round, red seal visible between the second and third square seal near the top is too faded to be read.

Near the 11 o'clock position, just to the left of the first square red seal, is a vertical line of five bold, dark ideograms. This is the signature of "Colonel Nagakubo". Dropping almost straight down, near the 7-8 o'clock position in front of the tiger's mouth, is one of the few slogans present on this flag; most of the lines of *kanji* bearing the names of well-wishers. The uppermost ideogram stands out like a letter "t". The *kanji* read "Seven Lives In Dedication To The Nation".

Tiger attacking measures 8.75" x 14" (22.23cm x 35.56cm).

Tiger Attacking Flag – 26" x 32.5" (66.04cm x 82.55cm).

This cream colored good luck flag has aged over the years and turned off-white in color. It has two original thin brown leather corner tabs sewn to the right-hand corners. The upper right-hand corner retains its white silk tie string, however, the lower right-hand string was lost. The fly and hoist side edges have been reinforced with added stitching as is typical. The bright red center orb, or *hinomaru,* was screened onto the white background field.

This flag is interesting in that it displays two different slogans in the hori-

Tiger attacking measures 2.25" x 5.25" (5.71cm x 13.34cm).

zontal 12 o'clock position. The five bold center characters are very neatly brushed onto the white field and say *Ki Buun Chokyu,* or "I Pray Your Military Fortunes Are Long Lasting". To the left and right of this prominent wartime saying, two additional *kanji* characters were placed. These two ideograms say *Hissho* and "Certain Victory!"

Painted in the lower right-hand corner is a colorful and elaborately detailed rendition of a tiger. Although the tiger is small, it makes up for its size in exceptional detail. The name and seal, just below the tiger's tail, is probably that of the artist, Hosui.

Tiger Attacking Flag - 26″ x 37.5″ (66.04cm x 95.25cm).

This wonderful good luck flag is made from silk and has aged over the years to a dark cream color. It has two brown mesh-canvas corner tabs sewn into each of the right-hand corners, both retaining their white silk tie strings. The hoist or "tie string" side of the flag has additional hem stitching to give the flag more strength should it have been knotted to a pole or rifle. The opposite, or fly side of the flag has added stitching as well in order to provide durability should the flag have been flown in the breeze. This flag is probably navy related as all of those signing seem to hold naval ranks.

There are three bold, vertical lines of *kanji* characters written on the right-hand side of the flag. The ideograms nearest the fly side edge say *Zo* Takifuji Iwao kun, or "Presented to Mr. Takifuji Iwao". To the left of that, the center line of characters gives the slogan *Teishin Hoko,* or "To Serve With One's Life". The third line of thinner *kanji* reads *Kaigun Gijutsu Tai-i,* Takeuchi Takashi, or "Naval Technical Captain, Takeuchi Takashi". The large dark character just below the previous third line is for the word *Dan,* which means, "Determination". Immediately to the left of that are two lines of characters that read, "Naval Technical Captain ,Nasu Nobuo". In the 5 o'clock position to the left is the name, "Captain Kaida".

There are a number of prominent vertical lines of ideograms in the upper left-hand aspect of the flag. The first line closest to the red sun says *Koshin No Tora,* or "Yellow Bodied Tiger". In Asian culture, the color yellow is associated with heroism, therefore, "tiger heroism". In the center, the thick black line of *kanji* characters read *Mo Ko Zokugo,* which translates as "Continuous Ferocious Attack!" The upper left-hand corner displays three vertical lines of thin characters with a 1" (2.54cm) square red seal just below. It is likely that these provide the personal information of the artist, due to their proximity to the tiger artwork. They say *Tora Doshi Tora Tsuki Umare,* or "Born On The Month Of The Tiger, Year Of The Tiger", T*oyokawashi Ushikubo* or "Toyokawa City, Ushikubo Town", and finally a name, Suzuki Hoichi. The red seal states Hoichi.

Toyokawashi is a city located in the eastern portion of Aichi Prefecture, and has access to Mikawa Bay. The Japanese built the Toyokawashi Naval Arsenal there in 1939 where it produced everything from machine guns, military issue swords and bayonets to binocular lenses and munitions. According to Eric Hammel's book *Air War Pacific Chronology: America's Air War Against Japan in East Asia and the Pacific 1941-1945,* thousands of volunteers and conscripts worked in the facility, including hundreds of middle school students and high school girls. In the closing days of the war, B-29 bombers carried out an air raid on the arsenal in Toyokawa. Over 2,000 people were killed during that mission.

Tiger attacking measures 12.25" x 9.25" (31.12cm x 23.5cm).

Wide World Photo dated 2.14.41. The caption reads: "Scene in a Japanese armament plant recently showing gun barrels in production... The Japanese are rushing production of all war materials and armaments."

Tiger Attacking Flag – 27.25″ x 32.5″ (69.21cm x 82.55cm).

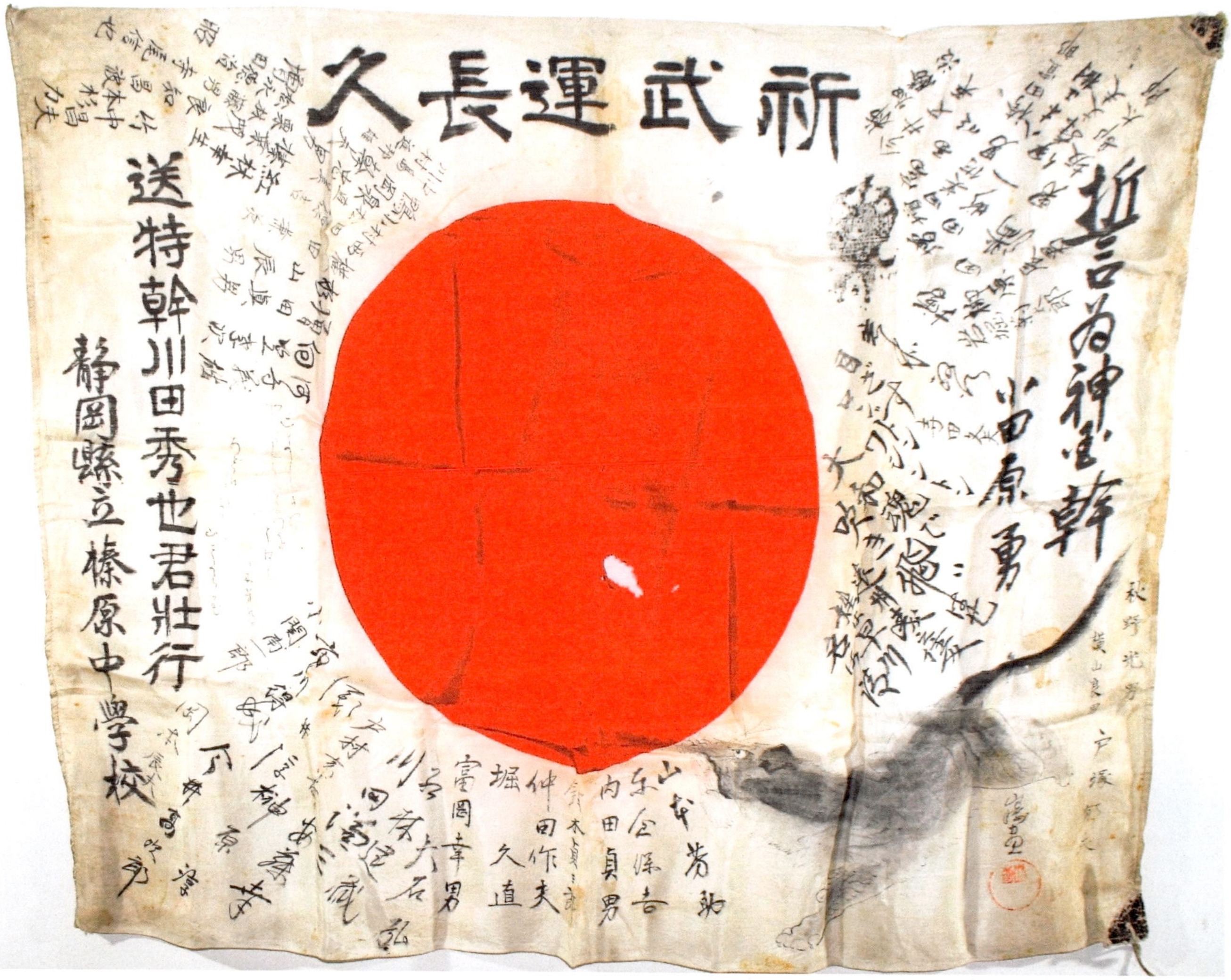

This white silk good luck flag is in generally nice condition, with a small tear in the center portion of the sun. The material has age toned to a cream color but is quite attractive. The two corner tabs are made from a thin canvas material covered with a thin layer of now peeling paint. The upper right-hand tab lacks its tie string, however, the lower tab has a small portion of its white silk string remaining. Typically, the hoist and fly sides of the flag show additional stitches placed along their edges for added strength. The deep red colored sun in the center is screen printed.

Placed horizontally across the top of the flag in the 12 o'clock position is the popular wartime slogan *Ki Buun Chokyu*, or "I Pray Your Military Fortunes Are Long Lasting". On the left-hand side of the flag are two vertical lines of medium sized characters. The first grouping closest to the red orb that contains ten *kanji*, gives the name of the man who received the flag, "Special Cadet Kawada Hidenari". The Special Cadet Program was begun in 1943 by the Imperial Japanese Army in order to train youth to be Non-Commissioned Officers. Immediately to the left of that, the second line with nine ideograms tells us that the flag was a gift from schoolmates at the "Shizuoka Prefecture Haibara Middle School". Shizuoka is located on the island of Honshu and was formed after combining three individual provinces: Totomi, Suruga and Izu. Tokugawa Ieyasu who was the founder and first of the Tokugawa Family shoguns, conquered the Shizuoka area, and later in 1868 the family established a residence there.

Below the tiger's rear legs, a small 1.25" (3.175cm) red seal and signature may be seen. This is probably the name of the artist who painted the illustration. Unfortunately, the signature is difficult to read. There is, however, an interesting exultation nearby that says, "Target London And Washington. Sweep Them Out With *Yamato* Spirit!"

The phrase, *Yamato-damashii*, or Japanese Spirit, dates back to the *Heian* Period (794 -1185), when Chinese cultural influences were predominant in many parts of Asia. Broad impact was seen from China in poetry and writing, language, art and business, to name just a few.

Tiger attacking measures 6.25" x 16" (15.87cm x 40.64cm).

A period of time ultimately emerged when outside influences were set aside, and Japanese ones were significantly embraced. Eventually, a separation between Chinese and Japanese-Spirit evolved. Beginning with the samurai in the Edo period (1603 A.D.-1868 A.D.), the phrase was seen in the context of *bushido* (loyalty and bravery). Later, the militarists and nationalists of the Showa time frame used this concept and combined it with *Shinto* in order to celebrate the essence of the Japanese people.

Tokugawa Ieyasu, (1543-1616)

Tiger Attacking Flag - 26" x 38" (66.04cm x 96.52cm).

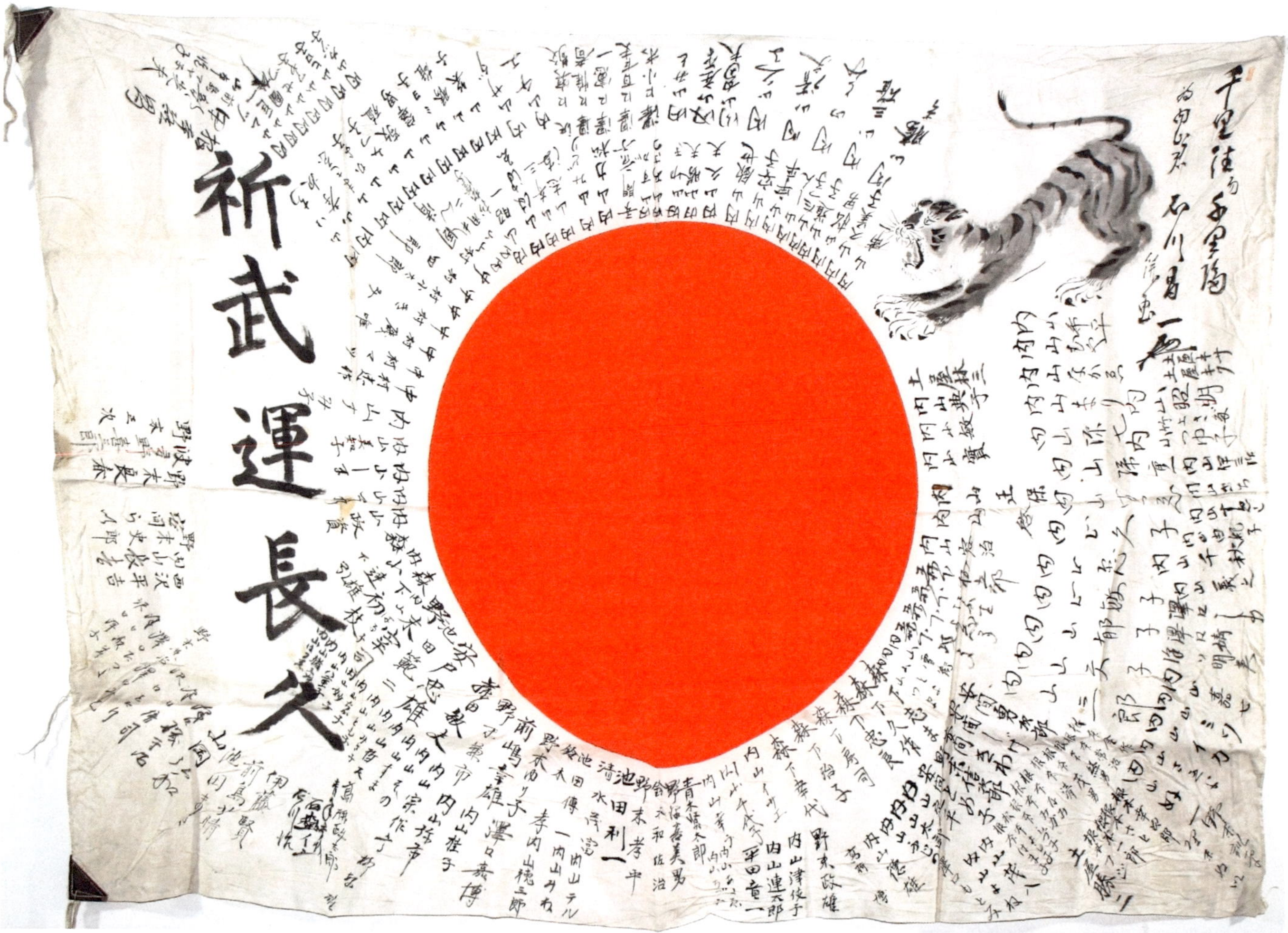

This white silk good luck flag has brown paper corner tabs that are made to look like leather. White silk tie strings are found in each of those two corners, and the hoist and fly side edges of the flag are hem stitched for strength. This flag is somewhat different to most; the corner tabs were placed on the left-hand side prior to signing. *Hinomaru yosegaki* were signed with the corner tabs facing in either direction. The vast majority reviewed, however, show the corner tabs and tie strings on the right-hand side. The red sun in the center of the white field was screen printed.

Tiger attacking measures 5" x 8.75" (12.7cm x 22.23cm).

Tiger Attacking Flag – 26.25" x 32.5" (66.68cm x 82.55cm).

This age toned silk good luck flag is in good condition with scattered areas of stain and insect damage. The right-hand side of the flag has both of its brown paper corner tabs sewn into place. The paper has been pressed in order to give it the appearance of pebbled leather. The upper corner tab has its long white silk tie strings securely in place, however, the lower string is missing from its tab. The hoist and fly side edges have been folded over and extra stitches added for strength. The bright red center orb was silk screened onto the white background.

Across the top of the flag there are five large sweeping black *kanji* characters which say *Ki Buun Chokyu,* or "I Pray Your Military Fortunes Are Long Lasting". Along the right-hand side of the flag is a large vertical row of seven ideograms. They give the name of the flag's owner and read *Zo* Hamaoka Kanzaburo Kun, or "Presented to Mr. Hamaoka Kanzaburo". To the right of that, nearest the hoist side of the flag, is another name, possibly the presenter, Hamaoka Masashi.

Of real interest, below the center character in the 12 o'clock position, a five-*sen* good luck coin may be seen sewn into place with black thread. A *sen* is a very small monetary unit, equal to one-hundredth of a *yen* (one *yen* equals approximately .009 U.S. dollars in 2020 terms). In the Japanese language, the number four, pronounced *shi,* has the same pronunciation as the Japanese word for "death". In the context of war, the number four may be seen to represent "the line of death"; a real possibility for any active duty soldier, sailor or airman in harm's way. Therefore, a good luck flag that displays a five-*sen* (five is pronounced *go-* in Japanese), bears the hope of going over or beyond the line of death that the number four represents. The meaning is that the person will successfully carry out his mission, ever mindful of his duty. Should death face him, it implies that he will move across that line without hesitation.

Near the 7 o'clock position is an interesting oval shaped red seal. The stamp measures approximately 1.5" x 0.5" (3.81cm x 1.27cm) and is perhaps a pen name, Tansui. Similarly, below the tiger's right rear paw, a signature and small rectangular shaped personal seal may be observed. The seal is too blurred to read but measures 0.5" x 0.25" (1.27cm x 0.64cm).

Tiger attacking measures 5.25" x 10.75" (13.34cm x 27.31cm).

Tiger Attacking Flag – 27.75" x 38" (70.49cm x 96.52cm).

This attacking tiger good luck flag is made of silk that was probably damaged and darkened from a long period of storage. Its two gold foil paper corner tabs, rather than being smooth, are textured to look like pebbled leather. Both corners have tie strings that are made from white cotton rather than silk. The hoist and fly side of the flag received added stitching for strength along those borders. The *hinomaru,* or bright red sun, in the center has been screen printed onto the white background.

There are only two vertical lines of medium sized *kanji* characters on the right-hand side of this flag. The upper line reads "May Your Military Fortunes Be Long Lasting, For Mr. Usuda Saburo". The lower line gives the name of the artist who painted the flag, Yano Kyoson.

Yano Kyoson (1890-1965), was an influential Japanese artist who was born in Ehime Prefecture on the island of Shikoku. As a radical *Nihon Nanga-in* member and exhibitor, he at one time studied under the painter Nakamatsu Shunyo. He was a regular presenter at the *Bunten* Exhibition (1907-1918), sponsored by the Japanese Ministry of Education Arts. For this major show, works of art were organized into three categories: Western Painting, Japanese Painting and Sculpture, with all three classes being awarded prizes. Because of its control over the selection of judges and the influence that those judges wielded, the *Bunten* had a long reach into the early development of modern art in Japan. The *Bunten* was replaced by the *Teiten* National Exhibition (1919), and after World War Two by the *Nitten* Exhibitions; Yano Kyoson exhibited in all three. He later went on to found the Osaka Art College.

Based upon the visible damage, it is probable that the subject flag was stored folded, along with a book or paper items (observe the very definitive lines of the stain), whose chemical composition was highly acidic. Over time, the difference in pH affected the silk fibers, and not only darkened them but weakened them as well. A 3" tear in the white field is evident near the 6 o'clock position, as is a smaller rip under the tiger's tail, possibly due to that exposure.

Timely research continues to be done at the laboratory in order to determine better ways to understand the causes behind the degradation of silk fibers, and how to preserve historically valuable items. Along with that work, more advanced methods of preservation will hopefully be forthcoming. Some interesting discoveries have been made recently, and those are discussed below.

We have long known that the basic structural component of degummed silk fiber is made up of the protein, fibroin. Before silk can be made into flags or anything else, the fibers must first undergo a degumming process. This procedure removes another protein, sericin, or silk gum. This is done in order to create among other things, an improved sheen (shininess) and texture to the final fabric. Studies show that the removal of the sericin leads to the inherent and eventual breakdown of fibroin. It may take many hundreds of years to finish the job, but as unsupported fibroin ages the pH drops (indicating increased acidity), and the silk begins to deteriorate.

This is important to understand because while outside components such as humidity, light and temperature, commonly understood to be the enemies of silk, do contribute to its degradation, there appears to be a natural breakdown anyway in the most basic segment of the fibers due to its normal process of aging. This is evidenced as a weakening of the individual fibers over time and may even be a contributing factor in the production of tears along fold lines in some flags. Similarly, folding and unfolding any silk material the same way, time and again, may cause the material to prematurely tear.

In their study sponsored by the Korean Research Foundation and supported by the Textile Conservation Centre, Kim *et al* concluded, after examining the damaged material, that museums may be jumping to premature conclusions in their treatment of silk fabrics. Rather than immediately assuming that the fabric was affected by outside factors, they need to be aware of the intrinsic ones that are occurring within the silk itself. They further point out that "Neutralizing the acid, for example by washing in a buffered [more neutral or basic] solution, may relieve concerns about acid-promoted peptide hydrolysis, [but it] will not repair the fragmented polymer..." and damage may occur.

Tiger attacking measures 18.25" x 6.5" (47.63cm x 16.51cm).

The artist Yano Kyoson, pictured in the front row, seated first on the left. Kyoson lost his left hand in an accident at the Osaka Artillery Arsenal.

Tiger Attacking Flag – 27.25" x 33.75" (69.22cm x 85.72cm).

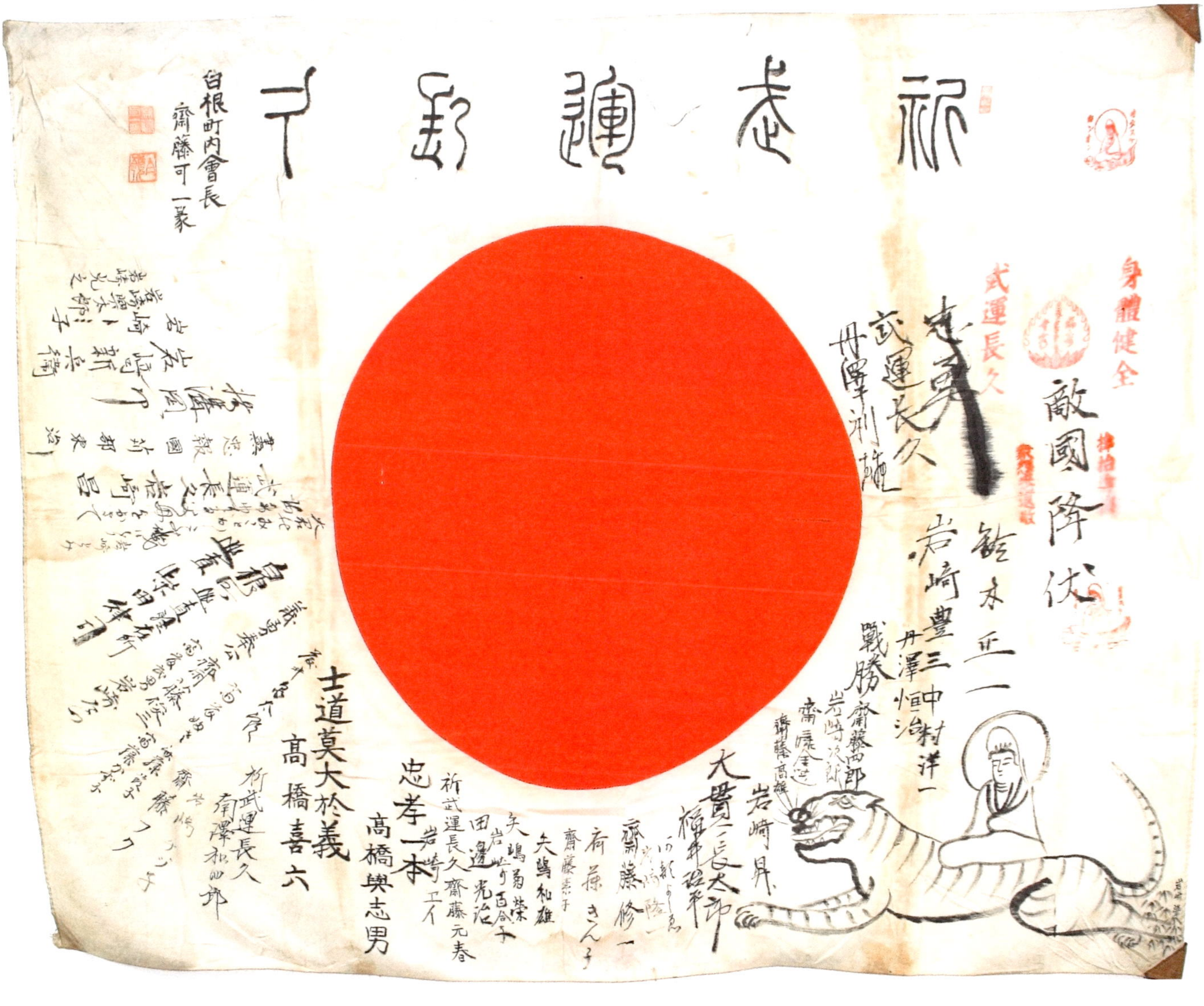

Tiger attacking measures 6" x 13" (15.24cm x 33.02cm).

The field of this silk good luck flag was originally a bright white color, but over time, has aged to a creamy shade of tan. There are two soft brown leather corner tabs sewn into place on the right-hand side of the flag. Both leather tabs have white silk tie strings in place that enable the owner to fasten the flag to a pole if desired. The hoist and fly sides of the flag (right and left-hand sides), have added stitching to provide strength to those high stress areas. The bright red sun in the center of the field was screen pressed onto the background.

Beside the tiger image in the lower right-hand corner, one of the most noticeable things about this flag is the five character horizontal slogan written across the uppermost aspect near the 12 o'clock position. The translation of the slogan is *Ki Buun Chokyu,* or "I Pray Your Military Fortunes Are Long Lasting". The ideograms stand out because they are done in a script that is rarely used any longer for this function. The slogan is written in an unusual t*ensho* style or script.

Different forms of script for the word *yama,* or "mountain", including *tensho* style.

The *tensho* style developed as an early form of highly decorative Chinese calligraphy practice during the *Zhou* dynasty (1045-256 B.C.). Later, the *Qin* (221-206 B.C.) variation of that script was adopted throughout China and became the standard for formal writing. It continued in wide use throughout the *Han* dynasty (206-220 A.D.), particularly being incorporated into name stamps, and was seen in other forms of stylistic engraving. The translation of the Chinese formal name, *Zhuansju,* or "decorative engraving script", accurately describes the ornamental form of Japanese *tensho*. Eventually, Chinese script evolved and with those changes rendered *tensho* little used in standard writing.

By the time *tensho* arrived in Japan during the *Nara* period (646-794 A.D.), it was no longer used in formal Chinese writing, and its ideograms were considered archaic even to their calligraphers. At that time, the Japanese used it more for decorative purposes as well, although their calligraphers still use it today. It continues to have ornamental uses, particularly in name seals/chops.

In the upper right-hand corner of the flag is an attractive red seal that measures 2." x 1.5" (5.08cm x 3.81cm). There are two rows of vertical *kanji* that each contains four characters. Those combine to say, *Otasuke* Kannon which means "Help From The God Of Compassion". This goddess is shown sitting upon a lotus blossom and is known in Japan as Kannon, also pronounced Kanzeon. Depending upon the situation, this deity may have either a male or female personification. In this instance she is known as the "Goddess of Compassion". This figure will be covered in greater detail later on in the chapter.

Located below the "deity seal" is a four character vertical line of red stamped *kanji*. When the characters are translated they say *Shintai Kenzen,* or "Good Health". The round temple seal immediately to the left of that is in the shape of a lotus blossom. Unfortunately, the ink is too blurry to read and translate accurately. To the left of the lotus seal is another vertical line of four *kanji* characters. Those are for the popular slogan *Buun Chokyu,* or "May Your Military Fortunes Be Long Lasting". To the left of that saying, three lines of vertical black *kanji* have been placed by the same person. The first line says *Chu-yu,* or "Loyalty And Bravery". The center line says *Buun Chokyu,* or "May Your Military Fortunes Be Long Lasting". The final line of ideograms, on the left-hand side, give the name of the well-wisher, Tanzawa Toshio. Just below the red lotus seal in the 6 o'clock position are four vertically written ideograms done in black ink. Those *kanji* say *Tekikoku Kofuku,* or "Enemy Countries Shall Surrender". Near the underside of the tiger's tail, partially obscured by bamboo is the name of the artist who painted the tiger, Iwasaki Hajime.

Moving clockwise past the tiger image, a number of name signatures are evident as are a few slogans. Between the 6-7 o'clock position, someone has written *Ki Buun Chokyu,* or "I Pray Your Military Fortunes Are Long Lasting," in a very fine thin line of characters near the edge of the red sun. Continuing from that point clockwise, four or five rows, you will next encounter two vertical lines of medium sized, thickly written *kanji.* In Japanese those say *Shido Go Wo Oite Bakudai, Takahashi Kiroku,* or "Righteousness Shall Be The Single Most Important Virtue Of The Warrior."

Finally, in the upper left-hand corner near the fly edge of the flag, two small square red seals may be seen along with two vertical lines of *kanji* characters immediately to their right. These say *Shirane Chokai Kaicho,* Saito Kaichiro Hitsu, or "Written By Shirane Town Committee Chief, Saito Kaichiro". Shirane is located in Asahi Ward, Yokohama, Kanagawa Prefecture.

The Buddhist deity shown riding upon the back of a tiger is known in Japan as *Kannon* or *Kanzeon.* In India she is called, *Avalokitesvala* ("The Lord Who Sees All Gazing Down") and in China she is customarily called *Guanshiyin.* Worship of this deity began in Japan in the late 6-7th century following the introduction there of Buddhism from Korea and China. Originally male, in some countries she now takes on a female form. Known as a *Bodhisattva*, or enlightened being, the deity is destined to become a Buddha but foregoes the bliss of nirvana in order to stay behind and help alleviate the suffering of those in the mortal world. According to the "Lotus Sutra", the body's form should be no obstacle to the pursuit of eliminating suffering. Thus, duality of form allows the deity to help anyone at any time. *(See appendix 5, page 192).*

Comparisons of this figure to a female deity were solidified in China sometime in the 12th century when a female Buddhist saint named Miao Shan was linked to *Guanshiyin.* It is said that the Chinese Princess Miao Shan, who lived around 700 B.C., wished to devote her life to the single joy of attaining Buddha-hood. Her father, the king, disagreed and ordered her to accept an arranged marriage. The princess refused her father's request saying, "I love not these pompous vanities; I swear that I prefer death to the so-called joys of this world." The king was so angered that he followed through with an earlier threat to execute her for disobedience.

At the appointed hour, the executioner took his sword and swung it down upon the Princess's neck; the sword was miraculously shattered. Next, an attempt was made to drive a spear through her body, but that weapon also splintered into pieces. Finally, the king ordered that his daughter be strangled with a silk cord. Knowing that this faithful and truly good girl was about to be sacrificed, Yu Huang the supreme heavenly being, ordered a messenger god known as the "Deity of the Place", to take the form of a tiger and hide among the crowd. At the exact moment of her death, the tiger attacked the group of executioners and scattered them about. He then placed the lifeless body of Miao Shan on his back and bounded into a nearby pine forest where he safely hid her body.

Once in the woods, the tiger was told to put a magic pill into the woman's mouth in order to preserve her body from decay, awaiting the time when her spirit would return from the lower regions and reanimate it. Prior to her death, the Princess had forgiven her executioner for the task that he was about to commit and accepted all of his lifelong negative karma upon herself. In this unfavorable condition, she was made to travel about the lower regions of hell, where she witnessed its many horrors. She became so upset at what she saw that out of compassion, she released all of the good karma that she had gathered to herself, allowing many trapped souls to escape their terrible fates. After this, hell became such a paradise that the King of that realm ordered her to be cast out and sent back to earth.

Immediately after she ascended back to earth and reanimated her body, Miao Shan was spirited away one more time by the tiger to a place called "Fragrant Island", where she was shown the path of spiritual perfection. On her journey to nirvana, however, she heard the cries of the destitute and suffering mortals of the earth. Looking back upon them, her heart was so moved that she chose to return to this realm as Guanshiyin, never to leave until all suffering had been assuaged. Today, she is represented as a beautiful young woman, dressed in white robes with a crown (*hokan*) upon her head and a string of prayer beads around her neck.

In Japan, this same Buddhist "Goddess of Mercy" is known in its appellation as Kannon. It is important to remember that Kannon, aka Kanzeon, has both male and female natures and may appear under different guises.

As Buddhism spread from India throughout the rest of Asia, the various deities began to evolve based upon those country's national traditions and/or customs. Arriving in Japan, they took on characteristics from a variety of different belief systems: Buddhism, *Shinto,* and Confucian practices to name a few. This system of combining numerous beliefs in order to form a more singular one is known as a syncretic belief system. To avoid any conflict in interpretation, it is simply posited that the deity has the ability to appear in whatever manner is necessary to alleviate suffering.

Kannon may also adopt the form of Bishamonten, or simply Bishamon. Known as the god of warriors, military men pray to him before battle in order to assure victory. The figure of Bishamon normally holds a lance in his left hand, and in his right hand, he holds a small pagoda that represents the treasures of Buddhist teaching.

The story of how Bishamonten arrived in Japan is a fascinating admixture of fact and legend. It is said that the "God of Warriors", Bishamonten, first appeared on Mount Shigi – located in the Nakoma Mountain Range of Nara, Japan – in response to

Statue of the deity, Avalokitesvala/Guanshiyin/Kannon.

the prayers of Prince Shotoku Taishi. The Prince came from the family of the ruling Soga Clan. He was a devout Buddhist scholar and early proponent of that practice at a time when *Shinto* was the traditional religion of Japan. In an effort to expand Buddhism throughout the country, the Prince found himself in a religious war of sorts against the powerful Mononobe Clan. As he prayed for victory over his opponents, it is said that Bishamonten appeared before him in the hour of the tiger, on the day of the tiger and in the year of the tiger. Afterward, he led the Prince's forces to victory!

The Prince eventually came to rule Japan from 594-622 A.D. and became known as the lay founder of Buddhism in that country as well as its first Buddhist statesman, transmitting its precepts nation-wide. Out of gratitude to the warrior deity for leading his triumph over Mononobe, the Prince built a temple in honor of Bishamonten on Mount Shigi in 594 and named it Chougosonshi-ji, also popularly known as Shigi-san.

Because Bishamonten led the Prince's forces to victory on the hour, day and year of the tiger, this third animal in the zodiac is given significant coverage as a guardian at the temple; its image is ubiquitous. Upon entering the grounds, visitors are greeted by the world's largest *papier-mâché* tiger – known as *Fukutora* – and have the opportunity, as they approach the site, to walk through the open mouth of a tiger who bestows happiness upon them.

Bishamonten, "Guardian of the Warrior" and defender of Buddhist teaching.

Prince Shotoku Taishi.

The tiger guardian at Chougosonshi-ji, Mount Shigi, Japan.

Tiger Attacking Flag - 26.75" x 37.5" (67.95cm x 95.25cm).

The Japanese good luck flag shown above is made from a thin silk material that has darkened due to the effects of aging. There are two gold foil paper corner tabs sewn into the right-hand corners, and each has two white silk tie strings that allow the flag to be attached to a pole or perhaps a rifle. The hoist, and fly sides of the flag have been typically re-enforced with additional stitches to add strength. The bright red sun in the center of the field has been screen pressed into its position.

This flag is unique not only in its artistically rendered tiger, but it has some unusual slogans and admonitions painted across the white field. The large characters placed horizontally across the top of the flag (12 o'clock position) say *Seishin Itto,* or "Unwavering Spirit!" To the right of that, just below the attacking tiger's paws, is a small block of two lines of vertical characters accompanied by an artist's red seal that measures 1" x 0.5" (2.54cm x 1.27cm).

While the ideograms and seal are difficult to read, they appear to say "By Ishida – 95 Years Old". Ishida appears to have been an accomplished artist, even at an advanced age, having captured the tiger in motion as it bursts forth from a bamboo thicket!

Approximately 5" below the artist's red seal are two long, thin vertical lines of ideograms. The message here is quite pointed: *Tatakai No Iyo Hageshiku Naru Toki E Mesa Ruru Kimi Yo Isamashiku Ike,* which translates as "The Battle Has Now Become Fierce, And In It You Have Been Called Upon. Go With Courage!" Moving clockwise to approximately the 6 o'clock position are two *kanji* characters that begin near the sun's edge and then move toward the white field. Those characters are for the popular slogan *Hissho,* or "Certain Victory!".

In the lower left-hand corner are four bold characters for the often written slogan *Buun Chokyu,* or "May Your Military Fortunes Be Long Lasting". This is normally seen written in a straight line either horizontally or vertically, but is sometimes encountered in any number of configurations, including *kanji* stacked upon each other as demonstrated here. Above the slogan just discussed, in the upper left-hand corner are two long vertical lines of ideograms; the one along the outer fly edge is thinner and not as bold as the one immediately to its right. The thinner line is for a type of poem and reads *Hana Wa Sakuragi Hito Wabushi,* or "Cherry Blossom Is The Zenith Of Flowers, And A Warrior Is the Zenith Of Men". The bold line to the right of that says *Dai Nippon Tei* [*Koku*] *Banzai*! This translates as "Long Live The Great Japanese Empire!"

Tiger attacking measures 9.5" x 9" (24.13cm x 22.86cm).

Tiger Attacking Flag - 26" x 40" (66.04cm x 101.6cm).

This beautifully illustrated white silk good luck flag is in average condition, although there is scattered insect damage on the left-hand side within the white field. The gold foil paper corner tabs are sewn into place on the left-hand side of the flag, while two white silk tie strings remain affixed to each. The hoist and fly side edges of this flag have added stitching in order to provide more strength to those high stress areas. The red sun has been screen pressed onto the center of the white field.

Just above the marvelous tiger illustration are three vertical lines of medium sized *kanji*. When combined, those characters say "The Nation's Rise Or Fall Shall Reside On Our Shoulders". The four large horizontal characters prominently displayed across the top of the flag read "Our Invincible Navy!" This navy reference is a likely indication that the owner of the flag was leaving home and entering the Japanese naval service. Seen in front of the tiger's right front paw are three ideograms that provide the name of the artist. They say "Drawn By Kanehiko".

The left-hand side of the flag has three rows of beautifully executed *kanji*. Unfortunately, this area of the flag has been damaged by insects due to improper storage and care. The five characters closest to the edge of the sun read *Ki Buun Chokyu*, or "I Pray Your Military Fortunes Are Long Lasting". The five bold lines of larger *kanji* immediately to the left of that present the flag to Mase Shinzo. The smaller line of vertical characters in the lower left-hand corner, accompanied by a square red seal say "Mase Steelworks".

Located in the upper left-hand corner is a large square red seal from the Prefectural Shrine in Wakayama Prefecture, Sasutahiko *Shinto* Shrine.

Mase Steelworks seal, 1" x 1" (2.54cm x 2.54cm).

Tiger attacking measures 12.25" x 12" (31.11cm x 30.48cm).

Wakayama Prefecture, on the Island of Honshu, is the site of the Sasutahiko *Shinto* Shrine. Directly to the north lie the cities of Osaka and Kobe. The area was once known for its thriving steel industry, but fell on hard times more recently when its largest manufacturer, Sumitomo Steel, moved most of its operations to China. In the center of Wakayama City lies Wakayama Castle.

At one time, the castle complex served as an operations center for the Saiga Ikki, a group of Buddhist warrior monks led by Saika Magoichi. The former castle was captured in 1585 following the Siege of Negoro-ji by the samurai warrior, Toyotomi Hideyoshi (1537-1598). He was able to divert the flow of nearby rivers that brought about the inundation of the structure, and forced the surrender of its inhabitants.

Wakayama Castle was rebuilt on the site of the former Ota Castle under the watchful eye of Toyotomi Hidenaga, Hideyoshi's brother. In a related note, the mountain where the castle resides is known as Mount Torafusa which means "A tiger leaning on its side".

Sasutahiko *Shinto* Shrine seal measures 2.25" x 2.25" (5.71cm x 5.71cm), Wakayama Prefecture.

View of Wakayama Castle from a hand-tinted postcard.

Portait of Toyotomi Hideyoshi.

Imperial Japanese good luck flag that belonged to Mase Shinzo, who was also presented with the tiger art flag previously illustrated.

This white silk *hinomaru yosegaki,* or good luck flag, belonged to the owner of the beautiful tiger art flag featured earlier. Japanese servicemen might be presented with more than one signed flag, with some never being carried into battle.

Many names are found scrolled across the white field of this flag. On the right-hand side, three separate rows of large vertical characters may be seen near the 3 o'clock position of the flag. Insect damage may be seen in this area as well, similar to that observed on the tiger flag, and probably caused by improper storage. The five characters closest to the flag's hoist side edge say, *Ki Buun Chokyu* or "I Pray Your Military Fortunes Are Long Lasting". To the left of that row, the next line gives the name of the person presenting the flag, Noda Nobuo. Finally, the third or longest collection of ideograms give the name of the flag's owner, "Presented to Mase Shinzo".

Tiger Attacking Flag – 28.25" x 38" (71.75cm x 96.52cm). *Flag Courtesy of Dean Kanabrocki*

This is a nicely made, white silk good luck flag that has age toned to a warm cream color. The colorful black, white, red, orange and brown painted tiger stands out boldly between the 5 and 6 o'clock positions. Below the animal's neck, a small horizontal line of characters probably identify the artist, Kurashina Jujiro. The fly and hoist side edges have been hem stitched for added strength and the brown paper corner tabs are still firmly attached by sewn threads. Long tie strips are correctly looped and knotted in each of the right-hand corners. The red sun, screen printed onto the center of the white field, has a small single blemish. In addition, there are some water stains scattered about the white background that do not detract from the flag's beauty.

Evident across the 12 o'clock position, along the flag's upper edge, are the five horizontally placed characters that say *Ki Buun Chokyu*, or "I Pray Your Military Fortunes Are Long Lasting". The first vertical line of three characters in the upper right-hand corner, closest to the hoist, say "Mr. Hara Kei" (or Katsura), probably the name of the flag's owner. Japanese names are sometimes pronuounced in a variety of ways, depending upon the individual's or family's practice. Just to the left of that is a long vertical line of smaller *kanji* that says, "Food Inspection Agency Chief, Okabe Masami". Positioned closely to the left of that in the 3 o'clock position, are two names that read Oshiro Kioshige and Shimoyama Shigenobu. Immediately to the left are two more vertical lines of dark, bold *kanji*. The first is an admonition that says "Bravery and Devotion. Spare No Pain! Head of Agriculture,

Aihara". This is followed by an added name, Ogawa Yasuaki.

In the lower right-hand corner, closest to the fly side edge, is a long, thin vertical line of characters. To the left of that are three smaller accompanying lines of ideograms. The first line says "Kanagawa Prefectural Food Inspection Agency, Fujisawa Branch" and then three names: Hirota Shigetada, Mihashi Kamekichi and a third name that I am unable to fully translate. It is likely that the three people named here were associated with the food inspection agency listed. In the 4-5 o'clock position just to the left of this are two larger lines of characters, angling up toward the lower border of the red sun. The first one says "Koza Kamakura Local Disease Prevention Office" followed the by the name, Tsutsumi Kinjiro.

Kanagawa prefecture is located in the Kanto region of Japan with its capital being Yokohama. Yokohama is the second largest city, by population, in Japan. It sits on Tokyo Bay and is an important deep water port. Following the *Meiji* Restoration in 1868, the harbor was enlarged to enable the expansion of Japan's silk trade with Great Britain. Kanagawa, part of that greater Tokyo metropolitan area, receives large numbers of visitors especially to the city of Kamakura. With a population of over 170,000 people, Kamakura is most noted for its ancient statue of Daibutsu, also known as "The Great Buddha of Kamakura". Cast in 1252, it is the second largest bronze Buddha statue in Japan, the largest being found in Nara's Todaiji Temple.

Fujisawa is a fairly large town that continues today in Kanagawa, while Koza is a district found within the central aspect of the prefecture. The area was once inhabited by some of Japan's earliest people groups, the *Jomon* (discussed in Ch. 1) and was mentioned in the *Nihon Shoki*, a book that chronicled the earliest history of Japan.

Tiger attacking measures 5" x 5.75" (12.7cm x 14.61cm).

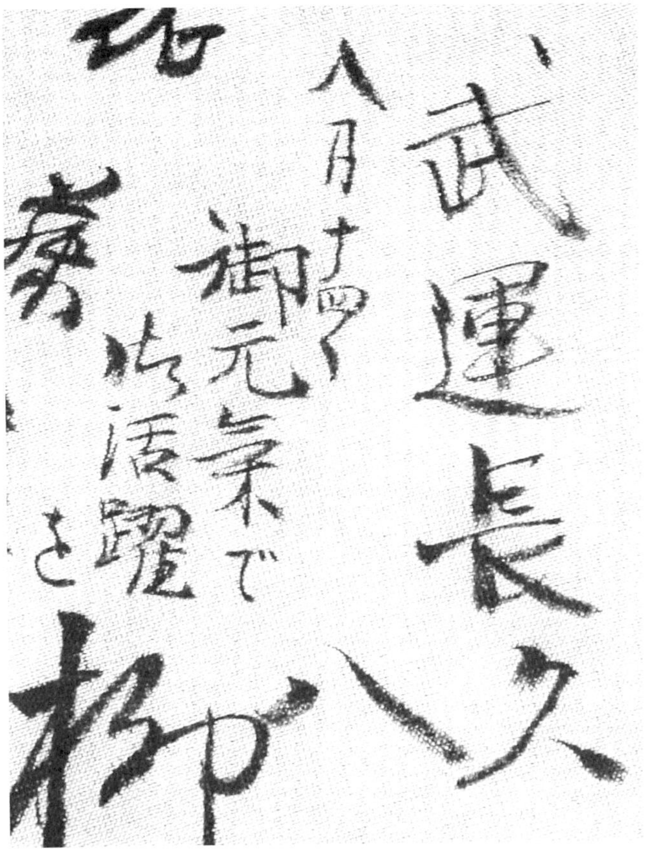

Finally, near the upper left-hand corner, are a few small lines of *kanji* characters. Beginning on the right, the ideograms read *Buun Chokyu,* or "May Your Military Fortunes Be Long Lasting". The adjacent lines read " Hoping For Success In Great Health- August 14th".

The statue of The Great Buddha, once covered in gold, represents Amida Buddha. The nearly 44' tall deity is shown being visited by sailors taking time off from a naval unit practice.

Tiger Attacking Flag – 26.25" x 30.5" (66.67cm x 77.47cm). *Flag Courtesy of Stephen Wallace.*

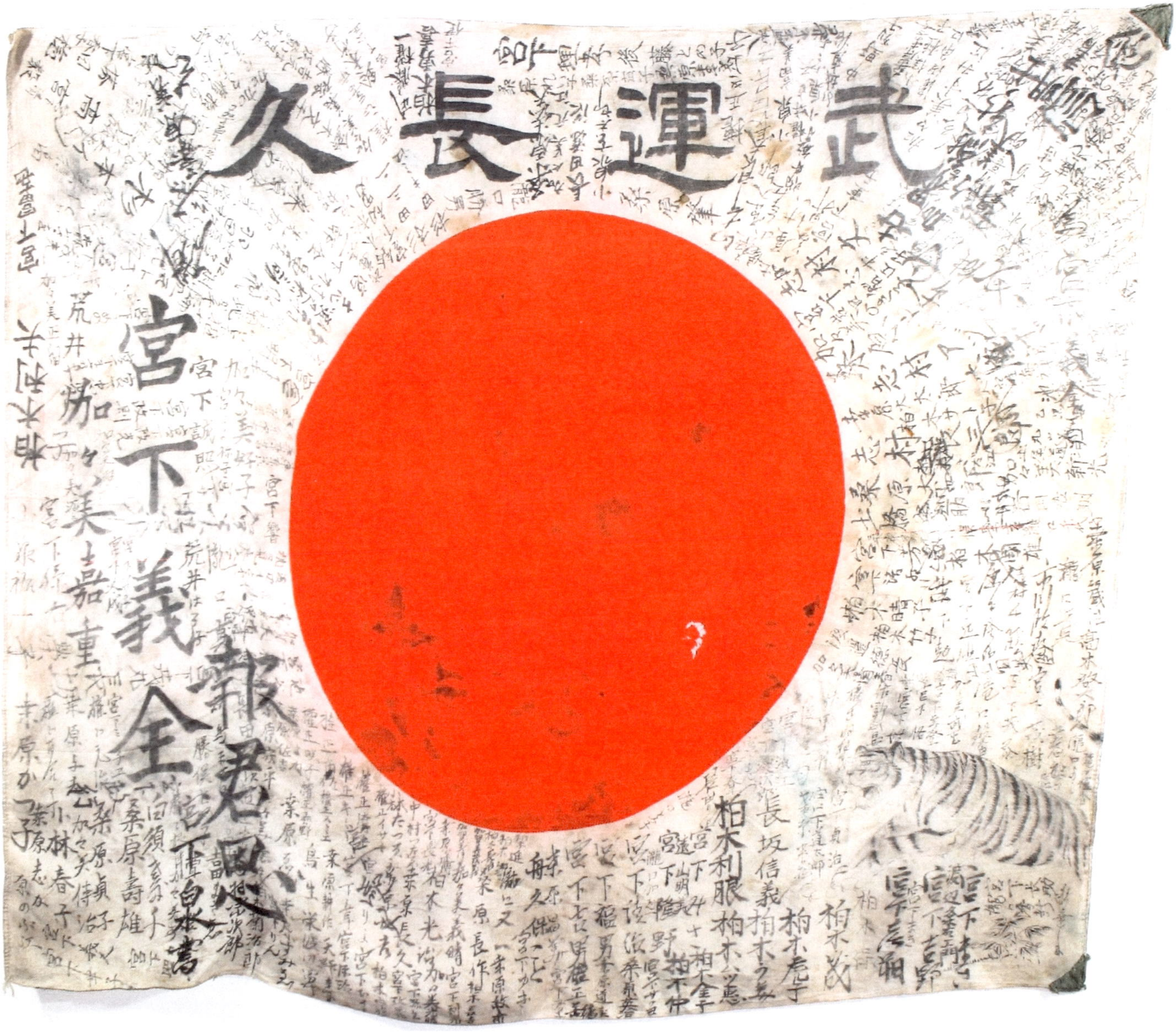

The Japanese good luck flag shown above is made from silk that at one time was white. As is common with this material, it has darkened with age. Compounding the color change is the extraordinary profusion of black signatures that fill the field; there are literally hundreds and hundreds of names added across the front of the flag, with many of them bearing the surname Miyashita. Both the hoist and fly side portion of the material have additional stitching added for strength.

The corner tabs are made from a green, gauze-type of material. The tie strings are no longer in place except for a very small portion on the upper right-hand tab. The dark red sun has been screen printed in the center of the white field and has a small tear present near the 4 o'clock position. From the 5 o'clock position, the extremelly detailed black and white tiger springs in the direction of the red *hinomaru* from just outside our field of vision. In one mighty leap, he clears the bamboo thicket below, seeking to neutralize his prey.

There are four prominent ideograms written horizontally across the top of the flag and they say *Buun Chokyu,* or "May Your Military Fortunes Be Long Lasting". In the far left-hand field, placed vertically between the 7-10 o'clock positions are three medium sized, bold vertical lines of *kanji* that really stand out from the lines of adjacent signatures. The group closest to the center orb says,"In Debt Of Gratitude To The Emperor". The other two vertical lines just to the left of that are more boldly written names from members of the Miyashita family.

Tiger attacking measures 4" x 8" (10.16cm x 20.32cm).

Tiger Attacking Flag - 30.5" x 37" (77.47cm x 93.98cm).

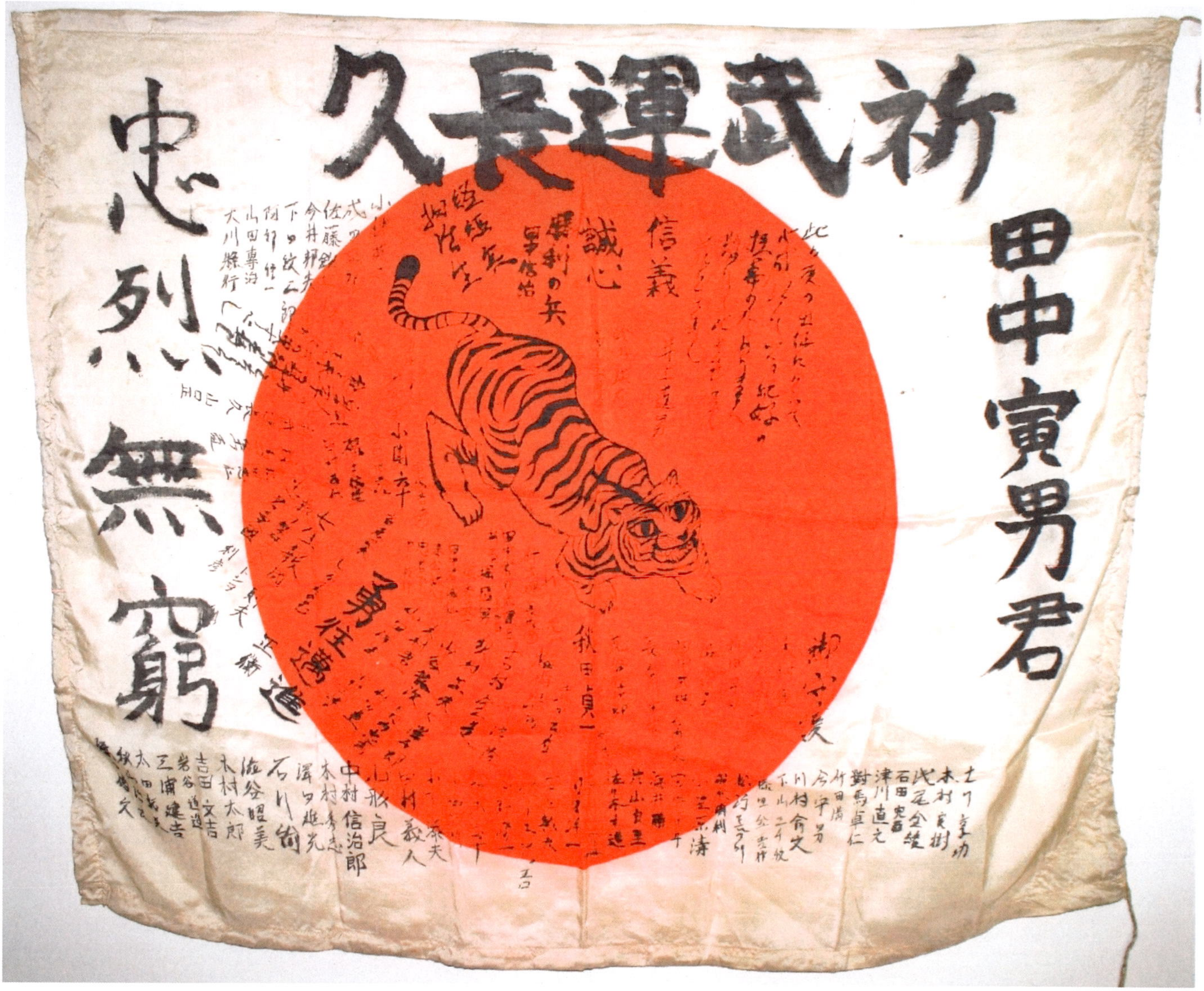

This extremely rare good luck signed flag is made from a heavy white silk. There are no leather or paper reinforcing tabs, however, the silk has been turned under and heavily sewn around all four corners and side edges. Heavy tie strings, or ropes, are present in both right-hand corners and appear to be made from cotton. The bright red sun was placed on the white field by a silk screening process.

The reason for this flag's rarity is the fact that rather than being a standard national flag, or *hata,* the artwork has been painted upon a flag normally issued to Japanese pilots for their emergency use. During the War, these were known as "pilot float flags".

All four sides of the flag have a 1" (2.54cm) thin hollow edge packed with a highly buoyant material that comes from the kapok tree (*Ceiba pentranda*). Found primarily in the rain forests of Central and South America, as well as parts of Southeast Asia, kapok is five times more buoyant than cork. The same material was also used in some life preservers as well.

Accompanying the flag was a hollow, collapsible black striped wooden pole containing an upper and lower loop that allowed the tie strings from the flag to be securely knotted to it. The flag would be folded and wrapped around the pole, then stuffed into a small pocket on the front of the pilot's float vest. In the event of a water emergency the pilot could pull the flag from the vest, extend the pole and use it to signal for help. The kapok would prevent the flag from absorbing water and sinking. In many years of collecting Japanese good luck signed flags, the author has only seen two pilot's float flags with painted artwork. This makes sense since it was possible that any float flag might eventually end up wet, damaging the inked signatures and any art that had been meticulously placed upon it.

Written across the top of this flag in the 12 o'clock position, five large, bold *kanji* characters may be seen. They say *Ki Buun Chokyu,* or "I Pray Your Military Fortunes Are Long Lasting". Directly to the right, positioned vertically along the hoist side edge of the flag is another bold line of five characters. This line provides the name of the flag's owner, Tanaka Torao Kun, or "Mr. Torao Tanaka". The tiger illustration positioned in the center of this flag is most appropriate, as the owner's name literally means "Tiger Man!" Moving across the flag to the opposite side is another vertical line of four large ideograms that say *Chu-retsu Mukyu,* or "Absolute Loyalty And Never Any Difficulties". Many signatures may be seen covering the flag.

Tiger attacking measures 10.25" x 15" (26.03cm x 38.1cm).

11. The Empire's Victorious Tigers

"Victory goes to the one who has no thought of himself."
– *Shinkage School of Swordsmanship*

It is said that Japan was founded by the Sun Goddess Amaterasu approximately 2,700 years ago. The nation's flag, a red sun disc in the center of a white field, represented the relational continuity passed down to all of Japan's rulers. Early history states that the first Japanese leader to display this style of flag was Emperor Monmu, sometime around 701.

Having ascended the Chrysanthemum throne on December 25, 1926, Emperor Hirohito's office far transcended merely being Japan's head of state; he was worshiped as a living god. As his country's sacred ruler, the Emperor was the inseparable embodiment of the nation; the sovereign and state were one. The flag was a highly visible, physical manifestation of this relationship between the Emperor and the "One Hundred Million", a phrase by which the Japanese people were referred at the time.

The style of flags that follow may be seen then as representing not only Japan, as it projected its colonial vision across Asia, but the tiger painted upon it further symbolizes the role of the Japanese warrior in securing that power afield. Known in Buddhism as one of "Three Senseless Creatures", the tiger symbolizes anger and might. The Japanese military was seen as the god given tool that enabled the Emperor and his leaders to stop imagining what the new Empire might be, and to act upon their long dreamed goals.

The Empire's Victorious Tigers Flag – 27.75" x 33.25" (70.48cm x 84.45cm).

Flag courtesy of Dean Kanabrocki

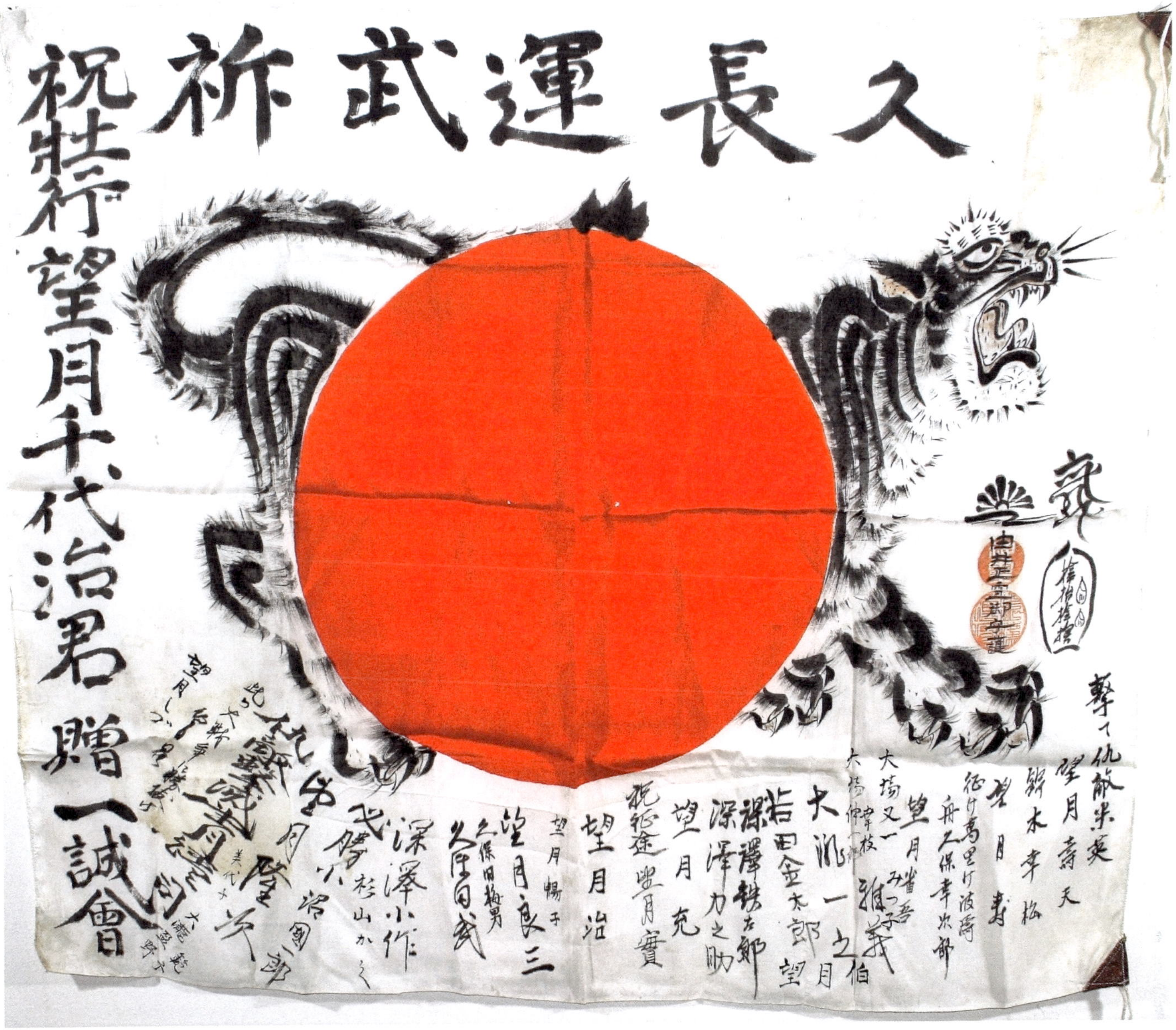

Incredibly beautiful is one way to describe this fanciful, full-bodied tiger that appears to be striking out from behind the red orb in the center of the flag. The heavy white silk material is in excellent-plus condition except for the dark stains seen in the upper right and lower left corners. Unfortunately, the material probably came into contact with another item that was highly acidic. The resulting stain is evident from the exposure. The two mesh-canvas corner tabs are well sewn into place and the brown paint that covers them remains undisturbed. Original white silk tie strings occupy their standard positions, and the hoist and fly side edges are reinforced with added stitches. As is typical, the red sun in the center of the white background appears to be silk screened.

Horizontally placed across the top of the flag are five massive *kanji* characters that say "I Pray Your Military Fortunes Are Long Lasting", or *Ki Buun Chokyu*. Of equal interest along the fly side edge is a long row of large bold *kanji* that offer a dedication. They say *Shuku Soko* Mochizuki Chiyoharu Kun, or "Celebration For A Send-Off For Mr. Mochizuki Chiyoharu". In addition, there are four large vertical characters in the lower left-hand corner that blend into the previous dedication. They provide the name of the person who gifted the flag and say *Zo* Isseika or "Presented By Isseika". Before moving to the right-hand side of the flag, there is an interesting slogan written between the 7-8 o'clock positions, just slightly to the left of the tiger's rear paw. The dark vertical row of mid-sized ideograms fan out in typical *hinomaru yosegaki,* or "sideways writing around the sun" and say *Kyuteki Gekimetsu,* or "Decimate The Enemy!"

On the hoist, or right-hand side, of the flag, just below the tiger's chin, are two colorful red seals as well as two small unique hand-painted images. Because of the elaborate fanciful script that covers them, the two red seals are most difficult to read.

On the right-hand side of the image is a rather large black and white hand drawn seal. The object, with what appears to be four *kanji* characters and two water drops running down its side, is in the shape of an urn. Measuring 3" x 1.75" (7.62cm x 4.44cm), the seal is quite prominent, even standing out in the presence of this amazingly detailed tiger.

Water is very important in *Shinto* beliefs as it represents purification. Cleanliness is symbolic of one's inner struggle to attain purity so that divine presence can be manifested. The Japanese have an expression for this, *nakaima o ikiru. Nakaima* means "in the now" and *ikiru* means "to live". *Shinto* teaches that rather than worry about past or future actions, each person should focus on what is at hand – the present. Awareness of the present is key to enlightenment. Presence can only be fully understood at the moment the person experiences it. The state of presence is said to be truth.

The Empire's victorious tiger measures 17" x 28.5" (43.18cm x 72.39cm).

A hand painted *Kikusui* (chrysanthemum and river), is seen in the upper left-hand corner, while a stylized water vase with the four characters for the mantra *Samuhara* may be seen in the lower right-hand corner.

In Japan, there is a mingling of *Shinto* and Buddhist beliefs to form a religious or philosophical melange. In Buddhism's case, there are specifically eight sacred symbols, or royal emblems, one of which is "The Sacred Vase". The vase holds "pure water", or "the water of life" which conveys compassion, wisdom, abundance and longevity. The washing away of impurities with this liquid is said to bring about enlightenment. Importantly, it is also said to be able to keep the forces of death away. *(See appendix 6, page 193)*.

Running vertically down the center of the water container are four small *kanji* characters. These are pronounced *Sa-Mu-Ha-Ra* and are an invocation to protect the bearer from harm in battle, particularly from bullets. The man who organized what became this new *Shinto* movement was Tanaka Tomisaburo (1868-1967). A former soldier who served in the Russo-Japanese War, Tanaka survived a harrowing, near death experience on the battlefield. He attributed his survival to a good luck amulet that he acquired that was marked with the four unrelated *kanji* characters that he pronounced *Samuhara*.

The *Samuhara* belief that developed during the Great East Asian War period was not an entirely new movement, but one that was originally established during the Edo period (1603-1868). Rather than evolving out of a strictly religious incubator where good luck amulets of various kinds *(omamori)* were sanctioned or normalized, the *Samuhara* belief was different. Its roots and early development were first established within popular conviction, that later morphed into common folk beliefs.

Throughout the Edo period, *Samuhara* grew as a general invocation against bad luck or injuries and even earthquakes. As the successes of the amulet were amplified, at least within popular belief, the *Samuhara* characters stamped upon paper became more plentiful. Fashionable essays written during this time, including one called the *Mimibukuro* introduced the use of the strange combination of *kanji* known more commonly in the day as *fuji* (a protector).

During the *Meiji* Period (1868-1912), Samuhara good luck charms containing the four characters became popular talismans to ward off evil in battle. For example, during the Sino-Japanese War (1894-1895) and the Russo-Japanese War (1904-1905) periods, the meaning of the amulet gradually narrowed until it became known as the one to carry in order to be covered from harm in battle from bullets.

Men returned from war and told stories about how their *senninbari,* also known as one-thousand stitched good luck articles, combined with the *Samuhara* characters, had saved them from certain death in battle. Over time the folk lore grew and became "truth" as part of popular culture. In spite of men being killed even while carrying the charm, stories persisted of miraculous survival with the common theme being the *Samuhara* amulet.

In 1935, Tanaka Tomisaburo worked to make the use of the Samuhara charm a common practice throughout Japan. He used his own funds to rebuild an old *Shinto* Shrine in Okayama and turned it into an operational center to make and distribute the amulet. In 1936, because the shrine was found to be unregistered with the local authorities, it was deemed to be a personal business enterprise and burned down. In spite of that, during the Great East Asian War, Tanaka continued to pass out his amulets. At War's end, he registered his movement, and re-built the shrine, later moving it to Osaka where it remains today.

Samuhara *jinja.*

The unusual hand painted emblem of the half-chrysanthemum flower floating upon a river, seen above the two red seals, is known as the *kikusui* or *kiku no mon*. The *kikusui* emblem comes from the family crest of the legendary samurai warrior, *Masashige Kusonoki* (1294-1336).

The story of Masashige Kusonoki's loyalty to Emperor Go-Daigo carried great symbolism among Japan's World War Two era military class. His forceful and cunning defense of Chihaya Castle in 1333 eventually led to the *Kemmu* Restoration of the Go-Daigo. A few years later, during the Battle at the Minato River, in which Masashige's forces were greatly outnumbered, this prominent leader chose to commit suicide rather than suffer defeat. This final act of loyalty to his Emperor cemented his courage within the pages of early Japanese military history.

The *Meiji* Restoration, which became official on January 3, 1868 marked the event after which Japan's militarists began to recognize the usefulness of connecting their country's early samurai lore to their plans for modernizing the country's military. This propaganda strategy had many facets that were utilized effectively during Japan's initial military successes and would continue to be implemented as an important tool right up until the end of the War.

Finally, below the *kikusui no mon* is a short vertical line of *kanji* written over the top of the two red seals that say, Yui Shosetsu *On Shugo*, or "Protected By Yui Shosetsu". Over the years, the story of Yui Shosetsu (1605-1651), took on minor folk lore status as a rallying cry of the common man; that story is sometimes related today.

Born as a commoner outside of the samurai class, Yui had outstanding leadership qualities and talents in the martial arts.

Because of that, he was mentored, and encouraged by a number of different *ronin* (unaffiliated samurai), who also instructed him in military strategy, sword play and other useful pursuits. He later became known as one of the "Three Great *Ronin*": Kumazawa Banzan and Yamago Soko being the other two.

Yui Shosetsu was so adept with the sword that he found work as a teacher at various academies. The instructional courses took into account more than the martial arts: they also taught history, philosophy and ethics, as well as politics. Some academies of the time were as much incubators of political thought as they were places to train, or develop physical skills and discipline.

There was no shortage of students in the academies, as policies under the Tokugawa shogunate led to the disaffection of many samurai. Students therefore came primarily from that class, and included many of the *ronin* who held grievances against the regime. They blamed the government primarily for loss of livelihood, loss of status, and loss of land/domain. To make matters worse, as new regulations were put into place, more samurai found themselves in a bind. Many more were forced into training at the academies.

Eventually, Yui Shosetsu opened his own academy and armorer's facility in Edo. As his success grew, so did his contacts with others of influence, including his fateful meeting with the skilled *samurai* Marubashi Chuya. Sometime after 1645, the two began to plot, along with a number of others, to overthrow the Tokugawa rulers.

Their goals were simple: they wanted to force the shogunate to stop confiscating the lands of *daimyos* at the time of their death, should they die without leaving any heirs. That practice disrupted the class structure and forced thousands of samurai (some say as many as 100,000 or more), into unenviable positions as master-less *ronin*. It further created issues for the government, as many of those disgruntled former samurai migrated to the larger cities looking for work.

Tokugawa Iemitsu (1604-1651).

Marubashi Chuya portrayed by the famous *kabuki* actor Ichikawa Sadanji, jr. Print by Natori Shunsen in the *Okubi-e* ("Big Head" style), 1931.

The plan had several phases. The stepping off point would be the death of the aged Tokugawa Iemitsu which finally occurred in 1651. Soon thereafter, Marubashi Chuya and his men were to use barrels of gunpowder to start a fire in Edo. A distraction of this magnitude was necessary to allow the group of rebels to break into the castle and eliminate several prominent government officials. At the same time, Yui Shosetsu and his men would seize key facilities in Sunpu (present day Shizuoka).

Unfortunately for the plotters, before the plot could be fully realized, Marubashi Chuya fell ill. In his delirium, he revealed aspects of the plan that led to the unraveling, and soon arrest of a number of the conspirators. Yui Shosetsu, and some of his men committed *seppuku* in Sunpu before the police could arrest them. Marubashi Chuya was placed under arrest, and executed in Edo. As was the custom, many of their family members were also taken into custody, and later be-headed or crucified.

The outcome of those events which came to be known as "The *Keian* Uprising", were beneficial for the samurai. Rather than run the risk of more violent clashes with the *ronin*, Tokugawa government officials were finally convinced to end the practice of escheatment or land confiscation. At the same time, they implemented programs designed to help the unattached samurai find employment.

The Empire's Victorious Tigers Flag - 27.5" x 33" (69.85cm x 83.82cm).

This is an example of a Japanese good luck flag that has seen some fairly heavy use over the years. The white silk is now heavily darkened with age and by grime. There are a number of small holes and tears present. Both corner tabs are missing. Each of the silk tie strings are still present, but no longer white. The upper tie continues to pass though its silk material, while the lower string is simply knotted around the material in the corner. The hoist and fly side edges have been strengthened with additional stitches of thread as is common in these national *hata*. The red *hinomaru*, or sun, was probably screen printed upon the white background.

The four bold characters written horizontally across the top of this *hinomaru yosegaki* say "May Your Military Fortunes Be Long Lasting", or *Buun Chokyu*. The vertical grouping of six large *kanji* placed on the left-hand, or fly side, of the flag identify the owner. The characters say "To Mr. Enomoto Hatsutaro". The vertical grouping of medium-sized ideograms that were placed directly in front of the tiger's head and paws are for the slogan *Gambare*. In English, *Gambare* means "Do Your Best".

The charging tiger, with ears back, mouth open and long fangs bared, is painted in a variety of colors that include: black, gray, white, red, orange-brown and gold. The artist has positioned the tiger toward the viewer. From that perspective, the animal's hind quarters appear small, as if in the distance, while the head, front legs and upper body are much bulkier, within reach.

The Empire's victorious tiger measures 10.75" x 24" (27.3cm x 60.96cm).

The Empire's Victorious Tigers Flag – 26" x 36.75" (66.04cm x 93.34cm).

This tiger art good luck flag is made from a white silk material that has age toned to a warm cream color. The brown paper corner tabs on the right-hand side are firmly sewn into place. The tab on the lower right-hand corner continues to sport remnants of its wartime civilian price control tag. Due to shortages, the Japanese government began to freeze prices by 1937 in an effort to avoid escalating demand costs and inflation. Many wartime manufactured civilian goods may still be found today with their original paper price control tags in place. The flat white silk tie strings are still visible and well in place in both corners. Typically, the hoist and fly side edges have been reinforced by added stitches. The red sun in the center of the white field has been screen printed.

Of note in the center of this flag (12 and 6 o'clock positions) are the large, prominent *kanji* characters for the popular wartime slogan *Ki Buun Chokyu,* or "I Pray Your Military Fortunes Are Long Lasting", (three ideograms above the sun and two below). Generally, when seen in its largest rendition, the slogan is written boldly in a horizontal fashion across the top of the flag (12 o'clock position) or may be seen written vertically in like style down the 3 and 9 o'clock positions.

Sometimes this particular slogan is observed with the characters split in varying ways; we noted this earlier on a number of flags. While this display was not typical, it is seen on enough flags to recognize that these are authentic, vintage World War Two Japanese *hinomaru yosegaki*.

The large vertical row of *kanji* characters nearest to the hoist side (right-hand) edge provide the dedication of the flag. They say "Celebration For Mr. Mizutani Bunichiro". In the lower center aspect of the flag between the 6 and 7 o'clock positions is a slogan that says *Uchite Yamamu,* or "To Attack And Decimate!"

Detail of the Empire's victorious tiger. The complete tiger art measures 20" x 28.75" (50.8cm x 73.02cm).

The Empire's Victorious Tigers Flag – 29" x 38.5" (73.66cm x 97.79cm).

This good luck flag has a large tiger painted horizontally across almost the entire front of the flag. The white silk material is age toned to a cream color, however, the black ink used to paint the big cat, as well as the many signatures, has rubbed off and stained the material. One brown paper corner tab remains in the upper left-hand corner, while the lower example has completely torn away. Unfortunately, neither of the two corners has its original tie strings present. The fly and hoist side edges of this good luck flag utilize additional stitches for strength. The round orb in the center of the white background was silk screened.

Across the top of the flag, the large *kanji* for *Ki Buun Chokyu,* or "I Pray Your Military Fortunes Are Long Lasting" are seen. Just below that slogan, a single large ideogram may be viewed near the sun's edge in the 1 o'clock position. The character says "Integrity". To the far right-hand, or fly side of the flag, six vertically placed medium-sized ideograms say, "Presented To Taisuke Shishido".

A number of Shishido Family members have signed the flag, and virtually all the other lines of *kanji* are signatures of those wishing good luck.

The wonderfully detailed tiger is quite large and done in ink shades of black, gray, white and red in the tongue/mouth/nose. Bursting forth from the bamboo thicket, the tiger's attack is reminiscent of the beast's stealth and guile as it silently pursues its prey.

The Empire's victorious tiger measures 22" x 30.5" (55.88cm x 77.47cm).

The Empire's Victorious Tigers Flag – 27.5" x 31.75" (69.85cm x 80.64cm).

Made from an age toned white silk, this Japanese good luck signed flag is in very good condition. Both gold foil paper corner tabs are sewn into place on the right-hand side of the flag, although each of them now appear to be more silver due to age. Two white silk tie strings are properly strung through the corners, and all are secure. The hoist and fly side edges of the flag have been provided added strength by hem stitching these heavier stress areas of the flag. The red sun in the center was probably screen printed onto the white background.

Located at the 12 o'clock position of this *hinomaru yosegaki,* two very large *kanji* characters say *Hissho* in Japanese, or "Certain Victory!" The large vertical line of four ideograms closest to the hoist or right-hand side of the flag say *Buun Chokyu* or "May Your Military Fortunes Be Long Lasting". Immediately to the left of that popular wartime slogan is the dedication to the flag's owner, "Mr. Kato Takahiro".

Just past the 6 o'clock position is a vertical line of five *kanji.* This is for the personal name "Setoguchi Masato". Boldly displayed in the lower left-hand corner, just in front of the snarling tiger, are four large vertical ideograms that say "All Of The Class Of The Tiger". The implication is that this flag is a going away gift to one of its members. The very last signature in the upper left-hand corner is that of "Hashimoto Shinya".

The Empire's victorious tiger measures 9.5" x 7.25" (24.13cm x 18.41cm).

The Empire's Victorious Tigers Flag – 27.25″ x 31.5″ (69.21cm x 80.01cm).

This is a Japanese signed good luck flag made from a white silk material that is now age toned to a cream color. In this example, the reinforcing tabs were made by adding a piece of triangular white silk sewn into each right-hand corner. Flat white silk tie strings were also stitched in between the layers of silk on both the upper and lower corners. Both hoist and fly side edges have received supplementary stitches. The flag's center orb was made using a less common two-piece construction method; the center of the white field was cut out to accommodate the red silk orb that was sewn into the void.

Written across the top of the flag in the 12-1 o'clock position are the *kanji* that give the name of the flag's owner. They say Inoue Takeo Kun, or [Presented to] "Mr. Inoue Takeo". The exuberant vertical characters and seal that were placed below the 3 o'clock position say Shigeru *Yori,* or "From Shigeru". This was probably the name of the artist.

The Empire's victorious tiger measures 16.5" x 28.5" (41.91cm x 72.39cm).

The Empire's Victorious Tigers Flag – 27.75" x 31.25" (70.48cm x 79.37cm).

This spectacular tiger painted silk good luck flag is a truly impressive work of art. The brown paper corner tabs are heavily sewn into their hoist side corners. Each of those retains its original thick white silk tie strings. The fly and hoist side edges have been given extra stitches during manufacture in order to provide added strength to the areas of the flag that experience stress when suspended. The red sun, or *hinomaru,* was screen printed onto the white field.

The three small, thin *kanji* located nearest the hoist side edge of the flag, immediately above the square red seal, say Go Iso which is probably the name of the person who illustrated the flag with the tiger's image. The red seal to the right of the tiger's head measures 1.25" (3.17cm) square and belongs to the above named artist.

To the left of the seal are four large vertical characters for the slogan *Buun Chokyu,* or "May Your Military Fortunes Be Long Lasting". The large, bold vertical lines of characters positioned on the flag's fly side edge give the name of its owner. They say *Shuku Nyuei,* Takada Hiroshi-kun, or "Celebrating Induction Of Mr. Takada Hiroshi".

In the illustration, the painter used black, gray, white, red, pink, light-orange, dark-orange and yellow-gold to portray in detail the animal, all the way down to individual strands of fur. Captured with a snarling growl, this prowling tiger pauses ever-watchful, looking back over its shoulder in an attempt to counter a hidden enemy.

Detail of the Empire's victorious tiger, which measures 13.75" x 22.5" (34.92cm x 57.15cm).

In this rare photograph, crew members of the Boeing B-29 *Bedroom Eyes* proudly hold up their war trophy; a Japanese tiger art good luck signed flag. The tiger and bamboo take up nearly 25 percent of the flag's white field.

Another photo of the crew of B-29 Superfortress *Bedroom Eyes* showing off their captured Japanese good luck flag. Note the full-bodied tiger near the upper left-hand corner. *(See appendix 7, page 193).*

The Empire's Victorious Tigers Flag – 27.75" x 39" (70.48cm x 99.06cm).

This over-sized silk good luck flag has toned with age and is now a light cream color. The paper corner tabs, khaki in color, are both present on the hoist side of the flag, as are the white tie strings. Hem stitching was added to the fly and hoist side edges in order to provide added strength to those areas. The larger than normal *hinomaru*, or red sun, in the center of the white field was silk screened to its background.

Two large *kanji* characters were boldly placed across the top of the flag near the 12 o'clock position. This slogan proclaims *Hissho,* or "Certain Victory!" The long vertical line of ideograms drawn nearest the hoist side edge give the name of the flag's owner, "For Mr. Ooba Fumio". To the left of this are two additional lines of vertical characters. The upper one is difficult to read as the writing is not very neat, however, it appears to say "Chief Of Plant Shimabara Matsuden". Directly below that is the name of the person who presented the flag, "Colonel Nagakubo".

The smallest vertical lines of characters written along the fly side edge of the flag say "Celebration For Advancement To The Field; The Tiger Of *Go-o*, Okada Ryo". Okada Ryo is the name of the artist that painted the tiger. The "Tiger of *Go-o*" refers to the practice of horary astrology. As practiced, an astrologer attempts to answer various questions by creating a horoscope for the exact time in which the question(s) was/were received and understood by the person constructing the chart. Answers may be formed as "Yes" or "No" responses, or may be more complex. The example here, in conjunction with an assortment of other signs, is intended to create the strongest Year of the Tiger. In this example, *Go-o* probably refers to strong leadership.

The Empire's victorious tiger measures 25" x 12" (63.5cm x 30.48cm).

The Empire's Victorious Tigers Flag 26.5" x 38" (67.31cm x 96.52cm).

Flag Courtesy of Vintage Productions.

Painted upon a fine white silk material, this lifelike tiger in shades of black, gray, white, red, pink and white, may be seen running out from behind the red sun in the center of the age-toned field. Both gold foil tabs are present in the right-hand corners and each has its tie strings present. The hoist and fly side edges of the flag were reinforced with added stitches for strength. It appears as though a portion of the fly-side edge was torn when the fabric became tangled on something. The orange-red center sun was screened onto the flag's background. Overall, this flag is in very nice condition.

On the right-hand side, near the 2-3 o'clock positions, the two largest vertical lines of *kanji* begin with the slogan *Nettoo Buun Chokyu,* or "Deep Prayers That Your Military Fortunes Will Be Long Lasting". Just to the left of that, the flag is identified to its recipient, Homma Masao. Immediately to the left of that are two smaller vertical lines of ideograms, one of them positioned above two red, square seals. The first line on the right says *Ki Buun Chokyu,* or "I Pray Your Military Fortunes Are Long Lasting. The line directly above the seals and the stamps themselves are too difficult to read.

The line of five *kanji* characters written horizontally across the top of the flag say *Banri Oorai* which means "Deep Prayers" in English. The smaller vertical lines written all across the bottom of the flag are the names of well-wishers, except for a single line that says,"Patriotism With Love".

Close up of the Empire's victorious tiger, which overall measures 5.25" x 30.75" (13.33cm x 78.1cm).

12. Tiger Associated Flags

"If you do not enter the tiger's cave, you will not catch its cub."
– *Japanese Idiom*

Tiger Associated: "Tiger Tail" Flag – 28.5" x 34.5" (72.398cm x 87.63cm).

The thin white silk of this tiger associated good luck flag has toned to a warm cream color. Unfortunately, the material is wearing thin in many areas due to age. The over-sized, heavy brown paper corner tabs have a pebbled finish in order to make them appear as real leather. There are two white silk tie strings in each of the right-hand corners. The right and left sides of the flag (hoist and fly) have added stitches to provide strength, although all along the fly-stitched edge the silk material has torn. The red center orb appears to have been screen printed onto the white field.

There are a number of ideograms written into the lower left-hand corner of the flag. The single large character on top says *Chu* and means "Loyalty". Directly below that are four characters for a personal name, Amano Naoyoshi.

Three long vertical lines of *kanji* are written on the right-hand side of the flag. The lines located on the right-hand side are difficult to read, but appear to say *Tame* Hama Takao-kun, or "For Mr. Hama Takao." Translated, the center line says, "I Pray Your Military Fortunes Are Long Lasting, And A Victorious Return!" The "tiger tail" style of script is a perfect accompaniment to the "Victorious Return" admonition. As a seasoned hunter, the tiger had similar qualities that were favored by the military. Besides being patient, brave and bold in the attack, the tiger was believed to have the ability to traverse many miles from home in search of prey, yet would successfully return victoriously from its mission. The very stylistic script on the left-hand column, closest to the sun, admonishes "To Leave And Serve As The Brave Shield." Unfortunately, the author is unable to determine the meaning of the wreath seen at the bottom of that line.

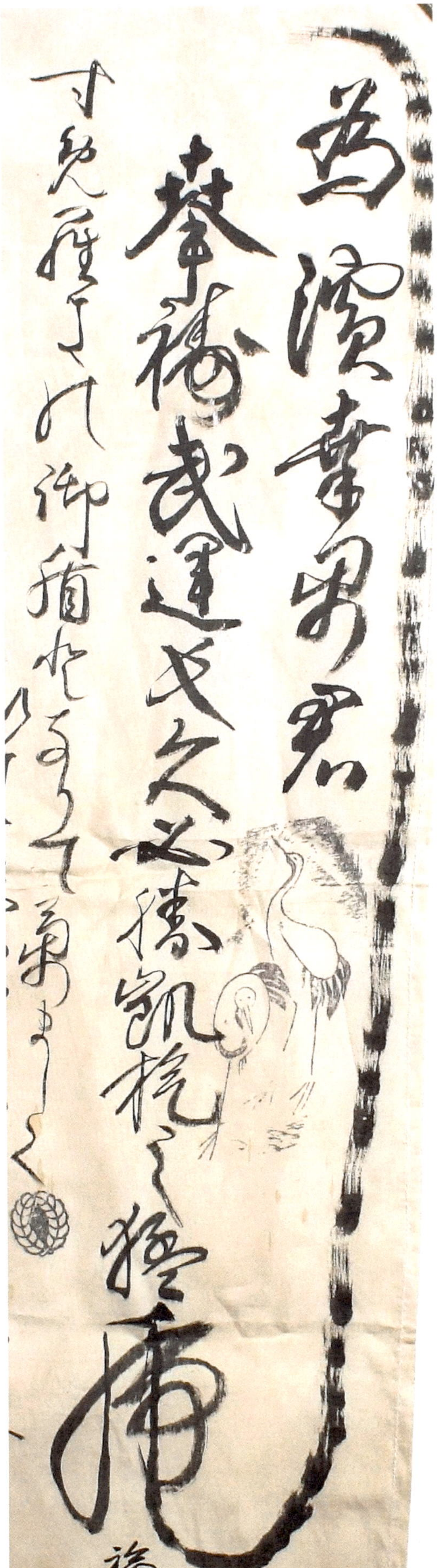

Stamped onto the flag near the hoist side edge in the 3 o'clock position are two cranes. One of the birds appears to be feeding, while its partner stands watch over the other bird. In Japanese culture it is believed that the fabled crane has the ability to live for one thousand years. Because of its long life, it is said to bring good luck and happiness. These are hopeful feelings intended for anyone facing grim prospects, especially someone going into battle.

Good luck "tiger tail" measures 23.5" (59.69cm). Two cranes, tallest measures 5" (12.7cm). Small wreath measures 1" (2.54cm).

Tiger Temple Seal Flag – 28" x 37" (71.12cm X 93.98cm).

This white silk good luck flag is in very good condition overall with only minor staining. The upper and lower right-hand corners have thick leather tabs firmly sewn into place. White silk tie strings accompany each of the stiff leather tabs, enabling it to be tied to a pole or rifle. Hoist and fly side edges have added stitching applied for strength. The red center orb has been screen printed onto the background field.

While this flag lacks tiger art, it has some interesting qualities and is tiger related. A bold *kanji* character may be seen in each of the flag's four corners. These combine to form the slogan "May Your Military Fortunes Be Long Lasting", or *Buun Chokyu.* It has been said that when the *kanji* for *Buun Chokyu* are distributed in this style around the red sun disc, it conveys the desire that the flag's owner will remember the full impact of the slogan, no matter where he serves across the Empire.

The observed placement of the four ideograms on this flag are a further indicator that this primary wartime send-off slogan, written profusely upon Japanese Good Luck Flags, had only general style requirements when added to each flag. Specifically *Buun Chokyu* and *Ki Buun Chokyu* were not drawn upon every flag and were done in a number of different ways. Sometimes the characters are thick and bold, while at other times the ideograms appear to be less enthusiastically (thinly) applied. In other examples, they may be seen applied in a straight horizontal line written right to left or left to right; a straight vertical line; as well as broken up in a variety of combinations or patterns.

The three vertical lines of *kanji* written over the two red temple seals proffer different blessings to the flag's owner.

The center line of characters from the temple flag says *Tora Yakushi* and represents the Yakushi Temple (*Yakushi-ji*), dedicated to Bhaisajyaguru, the "Buddha of Healing", on the day of the tiger. The Japanese also call him Yakushi Nyorai, or the doctor who cures suffering through the medicine of his teachings. *(See appendix 8, page 193).*

Yakushi-ji Temple Prayer: "At dawn I wake up seeing the head of a tiger – I see the light of the 'Lapis Lazuli Buddha' – How wonderful!"

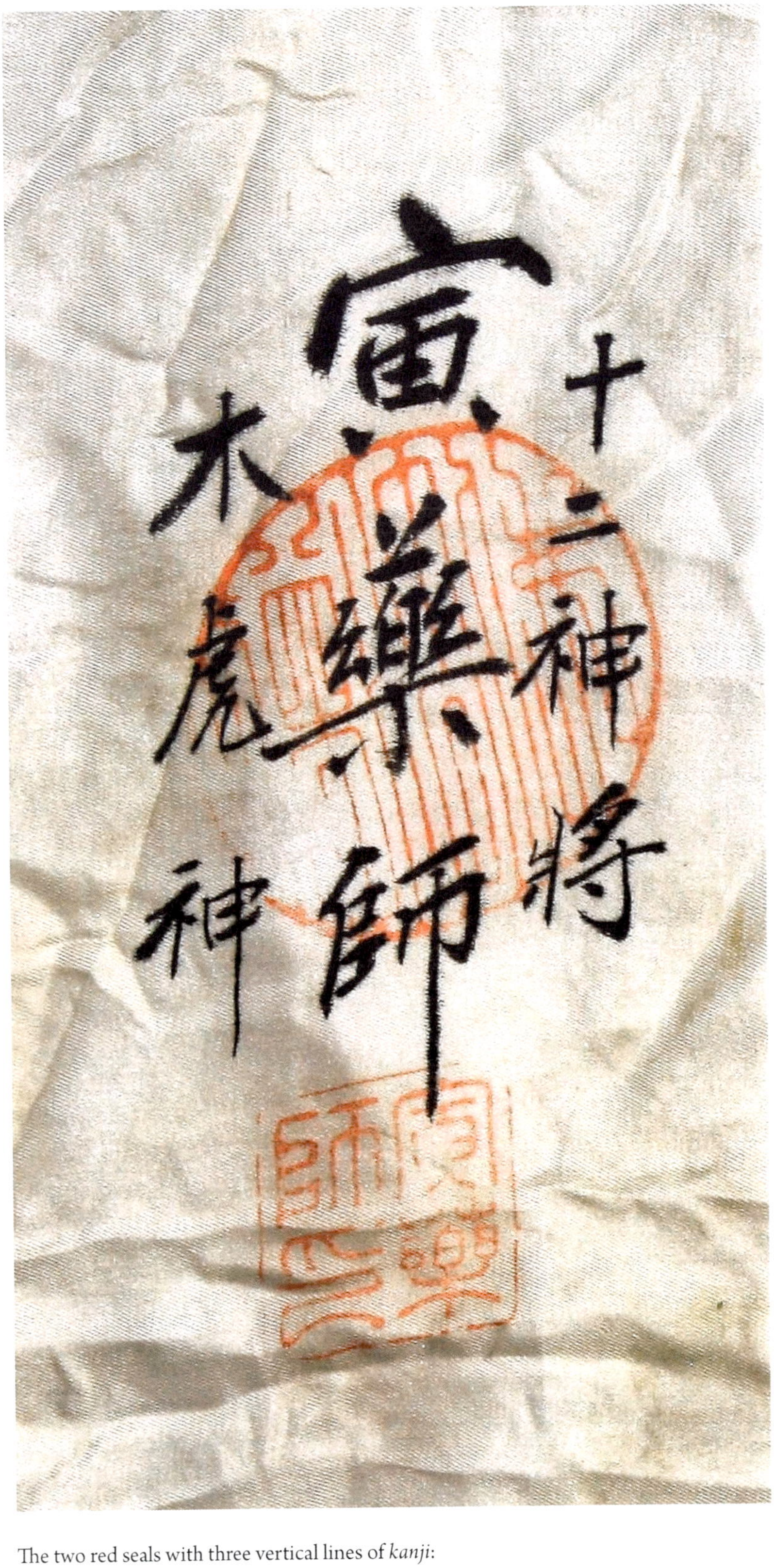

The two red seals with three vertical lines of *kanji*:
Upper round seal measures: 2" (5.08cm)
Lower square seal measures: 1" (2.54cm)

The famous Yakushi Imperial Temple, located in Nara, was one of the Seven Great Temples of Nara (*Nanto Shichi Daiji*) and is the center of the *Hosso* school of Buddhism.

Although not as prominent today as other schools of Buddhist thought, *Hosso* teaches a "consciousness-only" understanding in which reality comes from the individual's own mind and not through actual empirical experiences. They believe that the mind distorts reality, and projects it as reality itself. The "Buddha of Healing" was one of the first deities to arrive in Japan from China around 680 A.D.

The second vertical line of *kanji* characters located over the red seals on the right-hand side says *Ju Ni Shinsho,* or "The Twelve Divine Generals".

While some celebrate the "Medicine Buddha" in conjunction with a variety of other deities, when he is given a place of honor and placed at the center of devotion, he is normally seen surrounded by the figures of the "Twelve Divine Generals".

The Twelve were said to have been converted in their belief upon hearing the *Sutra* of the Yakushi Nyorai, and they now serve him. Whenever these specific Buddhist *sutras,* or scriptures, are put into practice by an individual or wherever human beings hold the "Medicine Buddha" in reverence and provide offerings to him, the "Twelve Divine Generals" will always be there to protect them. Those devotees will further be absolved from suffering and bad fortune as their deepest wishes will be satisfied.

The third vertical line of *kanji* characters, located on the left-hand side, says *Ki Tora Shin* or "The Wooden Tiger God".

The carved wooden statue of the tiger god, also known as the Tora Yakushi, is said to be the god of protection for individuals born in the year of the tiger, on the day of the tiger and in the hour of the tiger. The original statue located at the *Josen-ji* in Tokyo is said to have been carved by the well known *Shingon* monk, Kobo Daishi (774 -835 A.D.), who finished the statue on the day of the tiger and in the hour of the tiger.

According to legend, one day while the temple priest was yet a layman, he was taking a walk in the mountains when he was suddenly attacked by wolves. Desperate to escape, he sought protection by running to the nearby temple hall of *Hooray-ji.* Once there, he was saved by the Yakushi Nyorai (the "Buddha of Healing"), who had taken on the form of a tiger in order to defeat the wolves. The experience left the man so grateful that he took the vows of a Buddhist priest, and devoted himself to the Tora Yakushi. Thereafter, he placed a statue of that Buddha in a prominent place in his temple. Today, that statue is called the *Sainan Yoke No Tora Yakushi* or "Tiger Yakushi To Prevent Misfortune And Disaster".

Yakushi Temple in Nara

The carved wooden statue of the Tora Yakushi.

"Buddha of Healing", Bhaisajyaguru, from an 8th century woodcarving.

Ju Ni Shinsho "The Twelve Divine Generals." *(See appendix 9, page 192).*

13. Good Luck Cloth and Paper Talismans

"In the world of old memories, there's room for visitors."
– *Nobuhiro Watsuki*

Images of tigers were commonly seen in everyday, World War Two era Japan. Just as other cultures incorporate familiar or admired animals into various aspects of their lives, for hundreds of years the Japanese integrated the tiger into countless contexts that included: religion, literature, song, sports and the visual arts. Having earlier addressed tigers painted upon good luck flags, there was another military departure gift that, among other things, was sometimes decorated with black and white or color accented tigers. Normally meant to be commemorative/inspirational and bring good luck, these small pieces of decorated paper or cloth took a number of forms.

Small square, rectangular or irregularly shaped paper, silk or cotton cloth good luck amulets were frequently fashioned at home or send-off parties and presented to the men as gifts.

The dimensions of these items were generally much smaller than standard sized good luck flags, approximately the size of a white cotton handkerchief, and each was unique. Some were done on blank sheets of paper and looked like standard cloth Japanese good luck flags.

Paper *Hinomaru Yosegaki* Good Luck Flag – 19.29" X 13.77" (49cm X 35cm).

The paper good luck flag shown above is dated for the "Imperial Year 2600 (1940) by the Honjo Township Agricultural Association". Honjo is located within Saitama Prefecture, on the Island of Honshu. The large, bold slogan says *Hakko Ichiu*, or "I Shall Cover the Eight Directions and Make Them My Abode". This was a slogan first fully articulated during the Second-Sino Japanese War period and popularized through World War Two. It gained prominence in a speech that was given by the Japanese Prime Minister Fumimaro Konoe on January 8, 1940 and was part of a debate on whether to limit Japanese expansion in Asia.

The front of the amulet contains many signatures as well as a number of *tanka* poems. Some of the more interesting ones read: "Having looked up at the mountains after so many days, what a beautiful clear day in the midst of a monsoon"; "Rice paddy clapper slat-line has now turned into the national army line" (clapper lines are strings fitted with dangling wooden slats to fend off sparrows); and "The well that cools the watermelon may be dark but has a deep heart." Accompanying the signatures and poetry are comments like, "Your mother from home prays with tears at your brave figure".

Other examples were similar in style *(yosegaki)* to the good luck flag, minus the round sun in the center. These typically contained names, caricatures, slogans and even epithets reminiscent of the propaganda of the day. That being said, the Japanese could be masters of veiled sarcasm, with some entries being indelicately "pointed".

A typical celebration meal for a man entering service or going overseas. The group has presented their friend with a good luck amulet; a small piece of white cloth being held up by the man on the left-hand side. It consists of names, thoughts or best-wishes comments from those in attendance. Similar to a good luck flag, these *yosegaki* lack the round, red sun in the center.

Yosegaki-Style (Sideways Writing) Good Luck Cloth/Amulet with Caricatures.

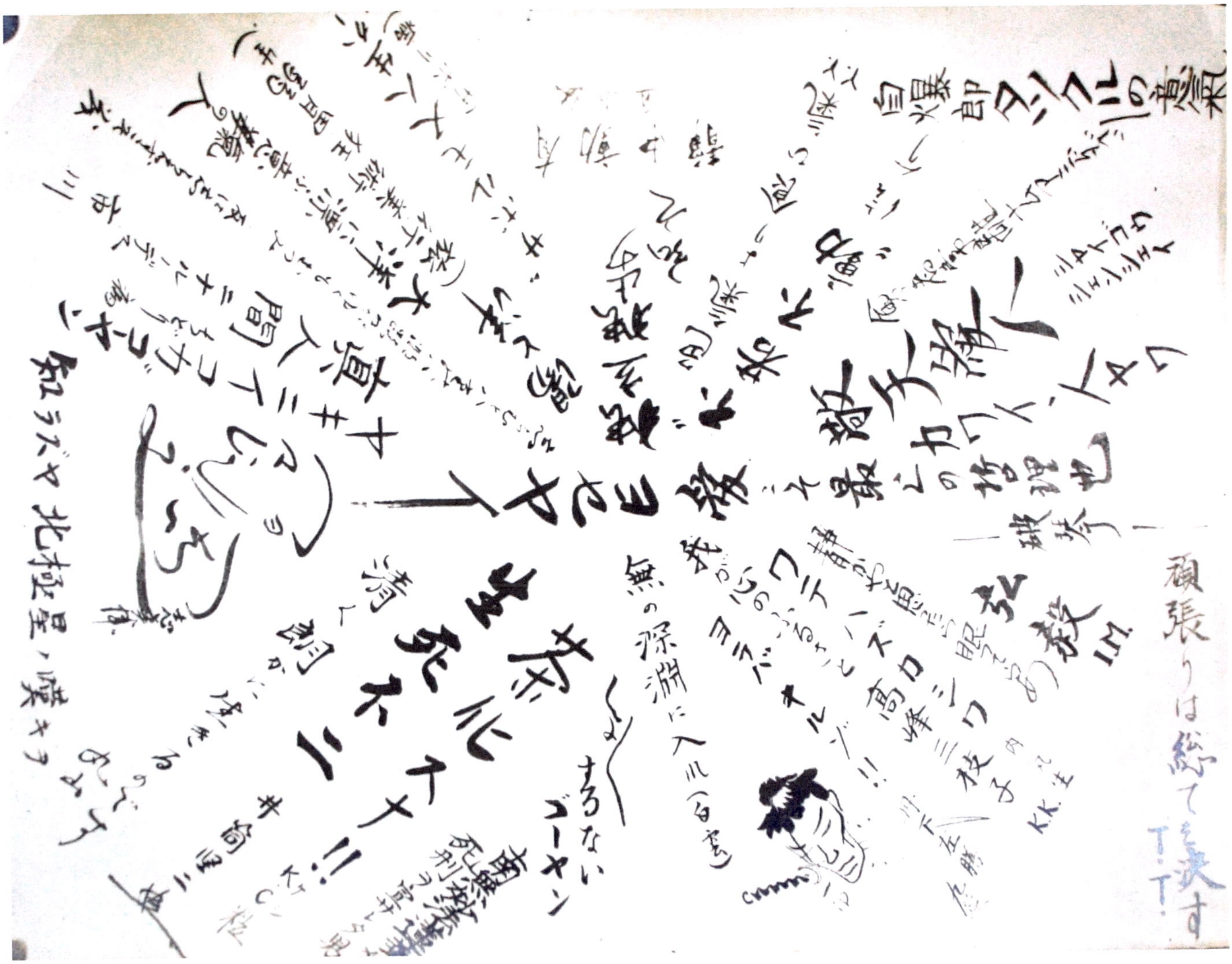

This *yosegaki*-style good luck cloth amulet is covered with signatures, caricatures and various phrases, written primarily using the Osaka-style dialect. The Osaka dialect comes from the Kansai region of Japan and is a coarser, yet at the same time, more poetic form of standard Japanese. Many of these phrases appear to be internal jokes among friends. Note some of the following: "To Be Lazy Is Also One's Life"; "Thought You Were Quiet, But Found Out You Were Just Sleeping!"; "Don't Do That!"; "Love Is The True Philosophy"; "Don't Fool With Me!"; and "I'm Embarrassed By This!"

In the lower right-hand corner, near the 5 o'clock position, someone has drawn an image of the head and sword of the one-eyed, one-armed samurai/*ronin,* Tange Sazen. This fictional Japanese print and movie character was made popular from 1927 through 1937. Known initially as the samurai Tange Samanosuke, he is ambushed and maimed by enemies that cut off his right arm and eye. Life for him loses all meaning following this betrayal, and thereafter he adopts the name "Tange Sazen". In the mid-1930s, the character was refined, giving up his nihilistic attitude to become a hero who fights against injustice. Tange Sazen is famous for the phrase, "I'll Cut You If You Come Near Me With Threat!" Reinvented and seen again in movies after the war, the character Tange Sazen continues to appear even today in videos and has been portrayed as both a male and female character.

Paper *Yosegaki* with Caricature Head and Face Created by Members of the Japanese 1st Company, 1st Ward.

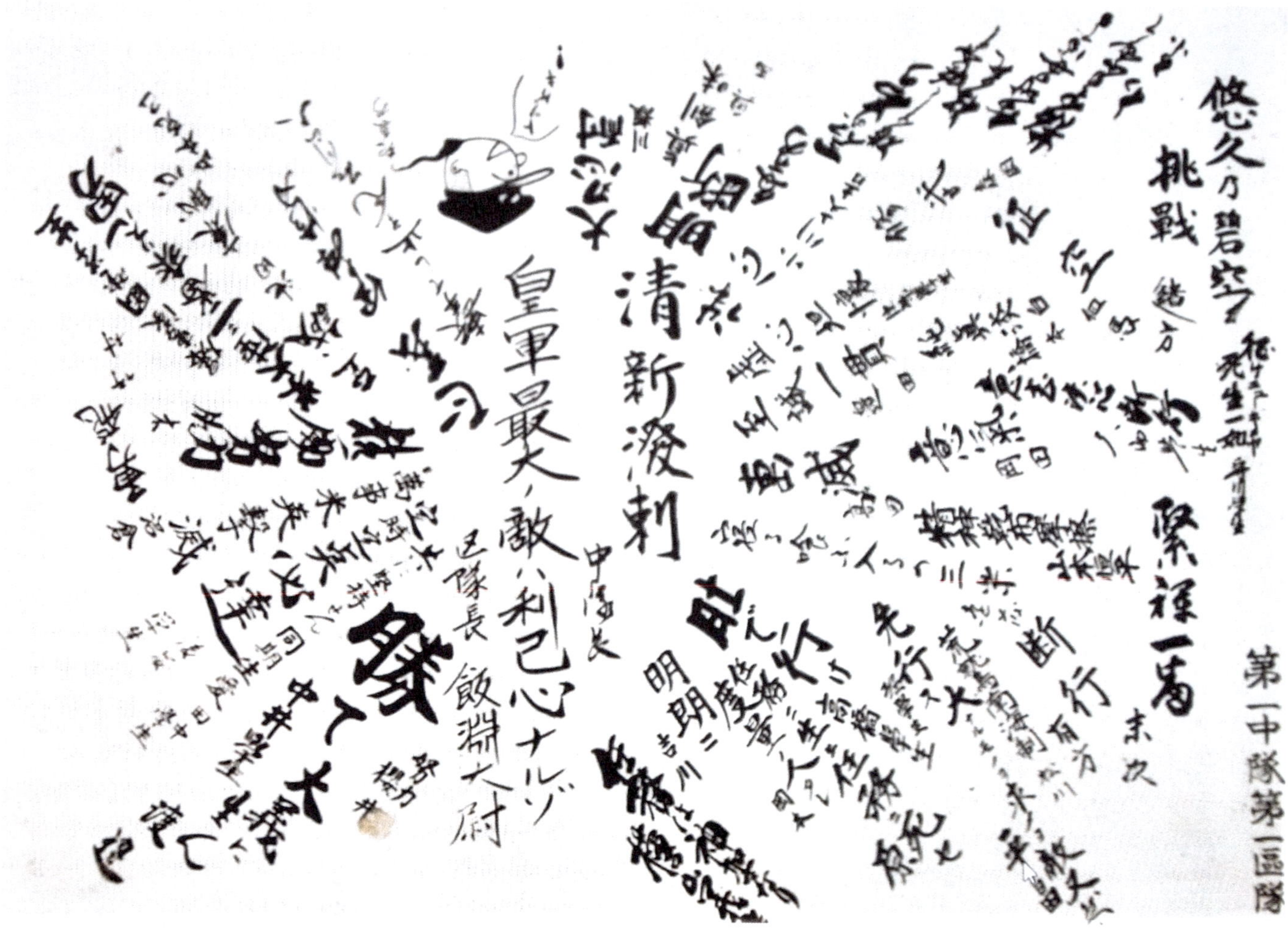

Japanese Army members from 1st Company, 1st Ward, creators of the *yosegaki* amulet above.

Unit members provided many strongly worded slogans, for example: "The Largest Enemy To The Imperial Army Is Selfishness. Signed by Ward Unit Leader Lt. Iibuchi"; "Be Pure And Energetic. [signed by] Company Leader"; "Off To New Guinea, Life And Death Is But One"; "Challenge The Great Blue Skies"; and others. The prominent caricature head sketched near the 11-12 o'clock position represents the main character from a Japanese cartoon known as *Fuku-chan*. Episodes by the well-known cartoonist Ryiuichi Yokoyama (1909-2001) ran from 1936 until 1971.

Yosegaki-Style (Sideways Writing) Good Luck Cloth/Amulet.

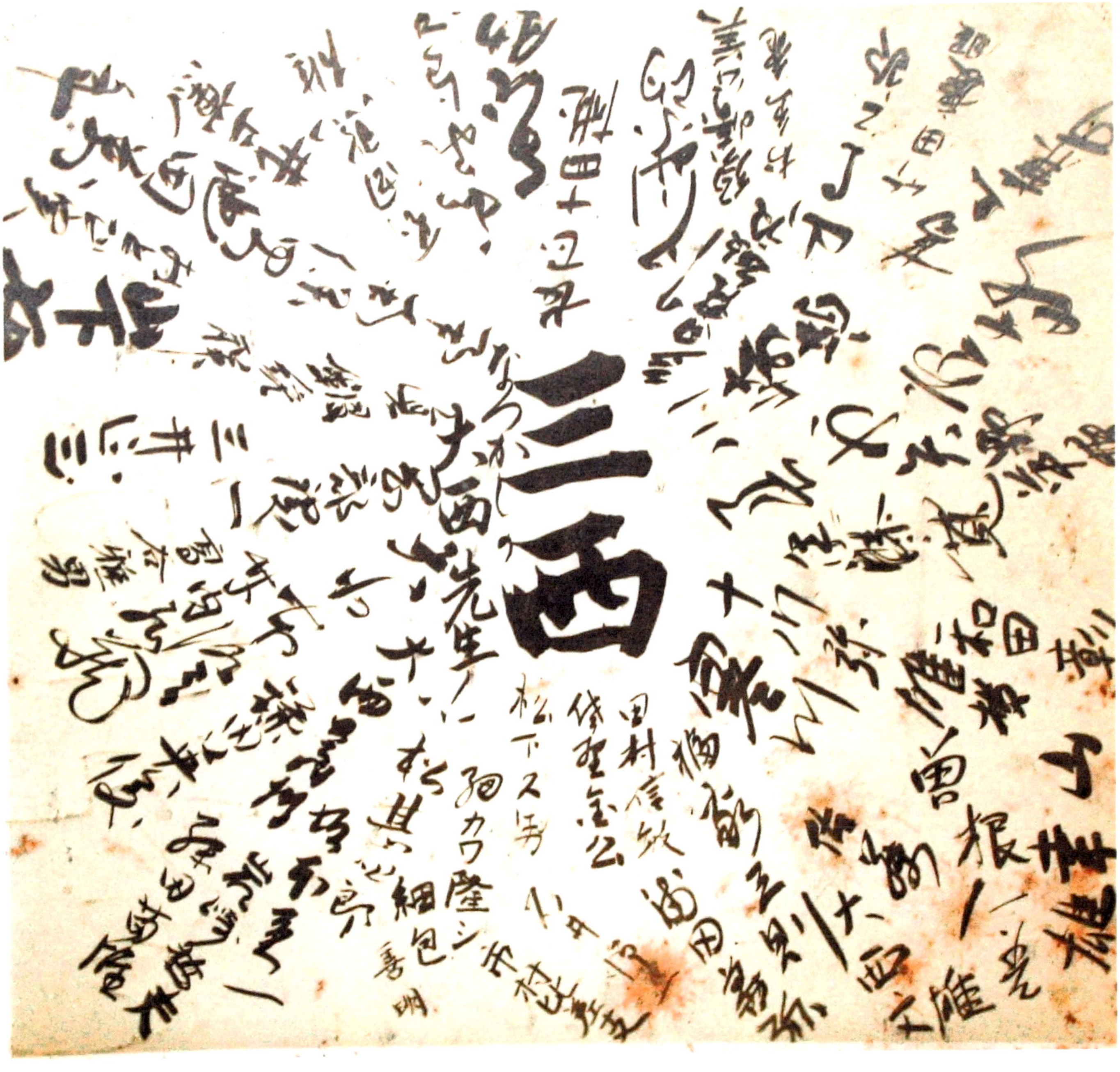

Covered with signatures that fan out from the center, similar to a good luck signed flag, this *yosegaki*-style good luck cloth amulet has a few prominent *kanji* characters placed directly in its center. The overall context is difficult to understand, however, the bold ideograms appear to read Mitsu-Nishi which may be a surname. Written in vertical characters immediately to the left of that appears to be a message that says"The Nostalgic *O-Nishi Sensei*", or "teacher".

An Unusual *Yosegaki* Good Luck Piece, Drawn on Top of a Map.

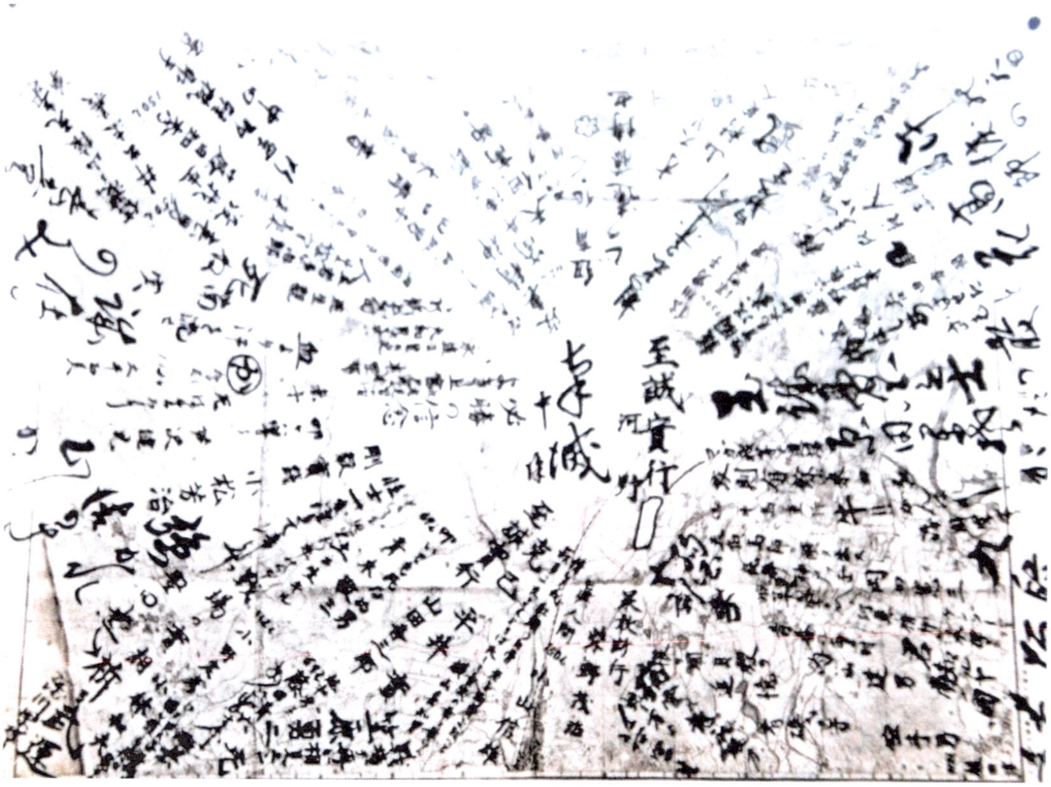

The front of the map has been signed with hundreds of names and contains numerous slogans. The two largest groups of *kanji* in the center circle say *Shi-sei Jikko,* or "Sincere Action" and *Geki-metsu,* or "Annihilate [The Enemy]". Other notable slogans include: *Kakan Dan-ko,* or "Absolute Determination"; *Tsuyokare,* or "To Be Strong!"; *Hissho no shinnen,* or "Absolute Victory Determination!"; and *Tataitemo shinanu otoko,* or "A Man Who Will Not Die!"

Yosegaki-Style Good Luck Cloth/Amulet - 15.5″ x 14.5″ (39.37cm x 36.83cm).

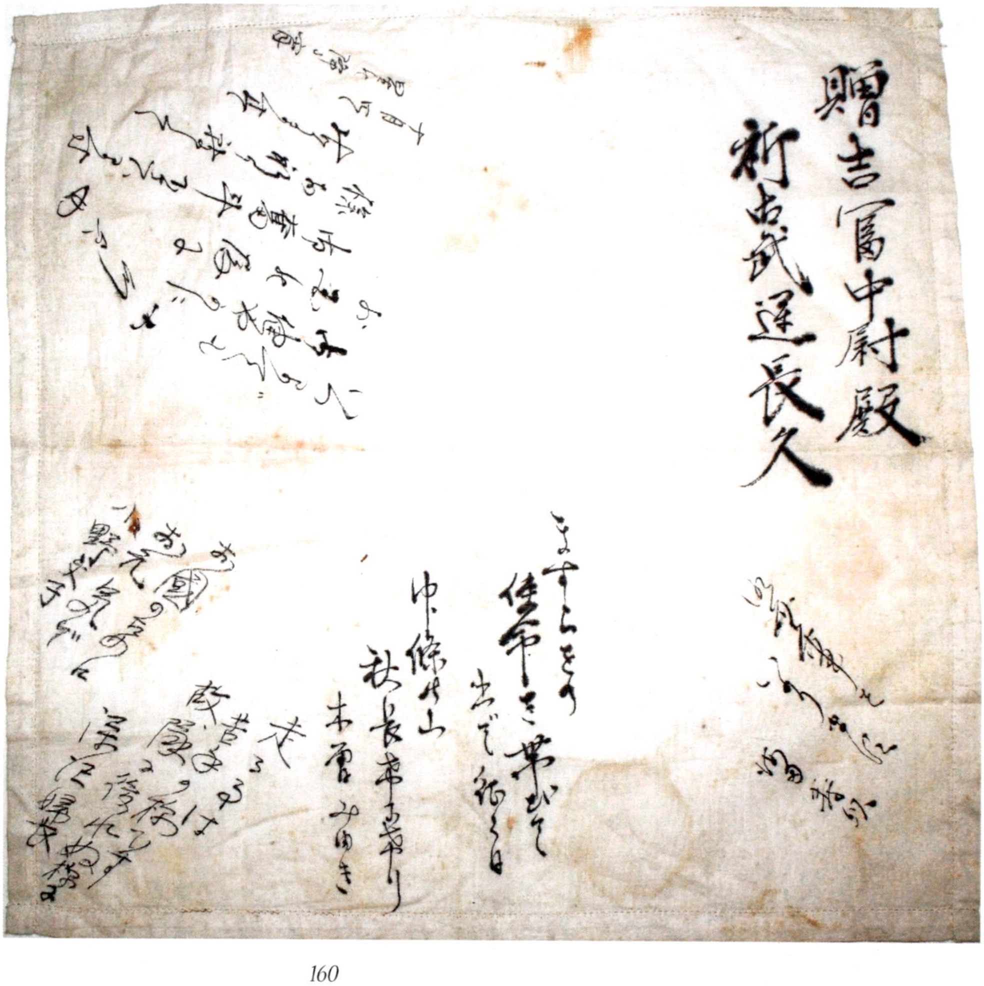

Signed on white cotton cloth that has now age-toned, this *yosegaki*-style good luck cloth amulet was presented to First Lieutenant Yoshitomi, whose name is shown in the upper right-hand corner nearest the outer edge. To the left of that dedication are the five vertical characters for the slogan *Ki Buun Chokyu,* or "I Pray Your Military Fortunes Are Long Lasting". Messages and poems written primarily by women may be seen inked across the face of the cloth, performed in a fanciful *soshotai* script. These are difficult to read but basically say, "Please Be Strong For The Country"; "I Will Pray For Your Health", etc.

Golden Kite Good Luck Cloth/Amulet - 12.25" x 12.5" (31.11cm x 31.75cm).
Hawks were used to symbolize a strong samurai spirit.

This good luck piece was meant to portray a dignified countenance. It illustrates a Golden Kite, glaring down from its perch in a pine tree. In Japanese mythology, the Golden Kite was a messenger of the *kami,* or gods, which was sent to provide aid to the future Emperor Jimmu in his fight against his enemy, Nagasunehiko. According to the ancient text known as the *Nihon Shoki,* or "The Chronicles of Japan", the two opposing forces met near Nara. During the ensuing battle, a Golden Kite swooped down and landed upon the bow of Jimmu. The large bird threw off a bright golden light that blinded and disheartened the opposition. In the end, Nagasunehiko would die and Jimmu would go on to become Japan's Emperor.

An unsigned woodblock print, "Hawk on Oak Tree", by the artist Kiyomasu, c.1710. Note the close similarities between the bird, here and on the amulet opposite: the pose, feather tufts on head, etc.

The pine tree, pine tree forest and mountain scene that comprise the subject matter in the artwork, along with the Golden Kite, provide a fascinating Japanese cultural narrative. Found among Japan's earliest animistic belief systems is the view that gods not only exist within, but also surround all things. As an example, an ancient *Shinto* narrative notes that at one time, a god alighted upon a large, old pine tree that was growing on a steep volcanic mountain and came to live there. Pine trees are known as *matsunoki* in Japan. When translated into English the word means "Waiting for a god's soul to descend from heaven". Distinct varieties of pine trees are believed to invite the souls, or *ogatama,* of the gods, and may therefore be planted at the entrances to certain shrines in order to invite them inside. The three *kanji* characters in the upper right-hand corner of this good luck art piece say, "Depiction Of A Glare".

Context, however, is a key to determine whether these examples were likely given as going away gifts to someone in the military or not. Those that follow were most probably military intended gifts. In that vein, it is important to reiterate that not every good luck flag accompanied its owner to war. In fact, there were probably some soldiers who received no good luck signed flag prior to departure, while it is known that other men were presented with more than one flag. Individuals normally only took one *hinomaru yosegaki* overseas. Likewise, some good luck cloth amulets may have accompanied their owners to war, while other examples remained at home securely packed away.

Storage space in the field was supposed to be limited to mostly necessary military related goods and supplies, therefore many unnecessary items or duplicates were left behind.

About to enter the military, Nakamura Kazumasa is photographed with his multiple good luck flags.

With the passing of the years, these small pieces of cloth, similar to their larger good luck signed flag "brothers", may have been forgotten, given away, or worse, been discarded by their owners as a bad memory of the war. Many have undoubtedly been lost forever. The examples on loan here were thankfully discovered by collectors and may now be shared with a new generation of admirers. For those who study and appreciate history, anthropology, art and other disciplines, these detailed and often stunning pieces of material culture are now highly prized.

For added impact, some cloth amulets had shrine seals stamped upon them, good luck coins sewn to their fronts, even *ofuda*, or paper good luck charms, may have been stitched to the cloth.

Tiger Good Luck Cloth/Amulet – 14" x 17" (35.56cm x 43.18cm).

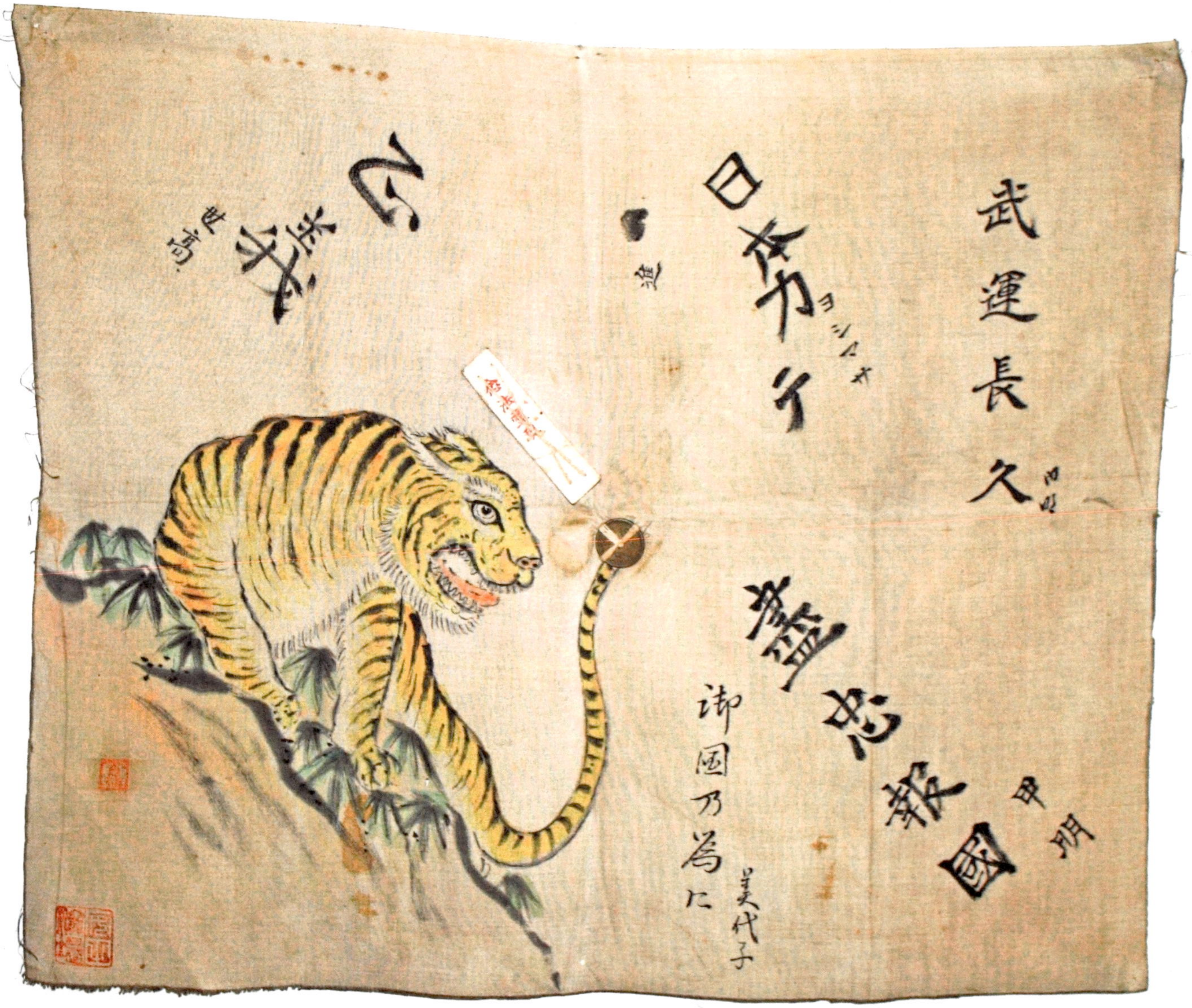

This beautiful tiger good luck amulet was fashioned from an unhemmed piece of white silk cloth. In the upper right-hand corner, the four vertical characters for *Buun Chokyu*, or "May Your Military Fortunes Be Long Lasting", may be seen.

To the left of that near the center, the characters for "Win Japan!" and "Justice" are seen. In the lower right-hand quadrant, in the 4 o'clock position are four bold ideograms that say *Jinchu Hokoku,* or "Serving The Nation With Utmost Loyalty". Near the lower-center, the smaller vertical line of Japanese characters say "For The Country".

A detailed tiger, crouching upon a rock amid a bamboo thicket, was painted in the lower left-hand quadrant by the artist using black, white, gray, red, orange, yellow and green ink or paint. At the tip of the tiger's tail, near the center of the cloth, a 5-*sen* good luck coin was stitched to the silk using white cotton thread.

In an unusual format, a white paper *ofuda* was stitched near the center-front of the cloth above the tiger's head. Its red characters say *Nenpo Reigen,* a prayer or slogan that was used by the Buddhist sect, rather new at the time, known as *Nenpo Shinkyo*. Established on August 3, 1925 by Reigen Ogura and still active today, it has by its own account over one million believers (511,000 adherents according to other sources), in Japan. According to the organization, "The spirit of *Nenpou Shinkyou* proclaims that all people throughout the world, regardless of religion, nationality or cultural background, should join hands in fellowship to build a world without conflict." *Nenpou Shinkyou* teaches the True Way in which we should live – the *Goseikun,* or five precepts. Its main temple, known as *Kongouji,* is located in Osaka, Japan.

Above: Two small red seals near the edge of the material in the left-hand corner. The upper seal is quite blurry, while the characters of the lower example are too highly stylized to read. Upper red seal measures: 0.25" square (0.64cm square). Lower red seal measures: 0.75" square (1.9cm square).

Right: Paper *ofuda* measures 0.5" x 2.5" (1.27cm x 6.35cm).

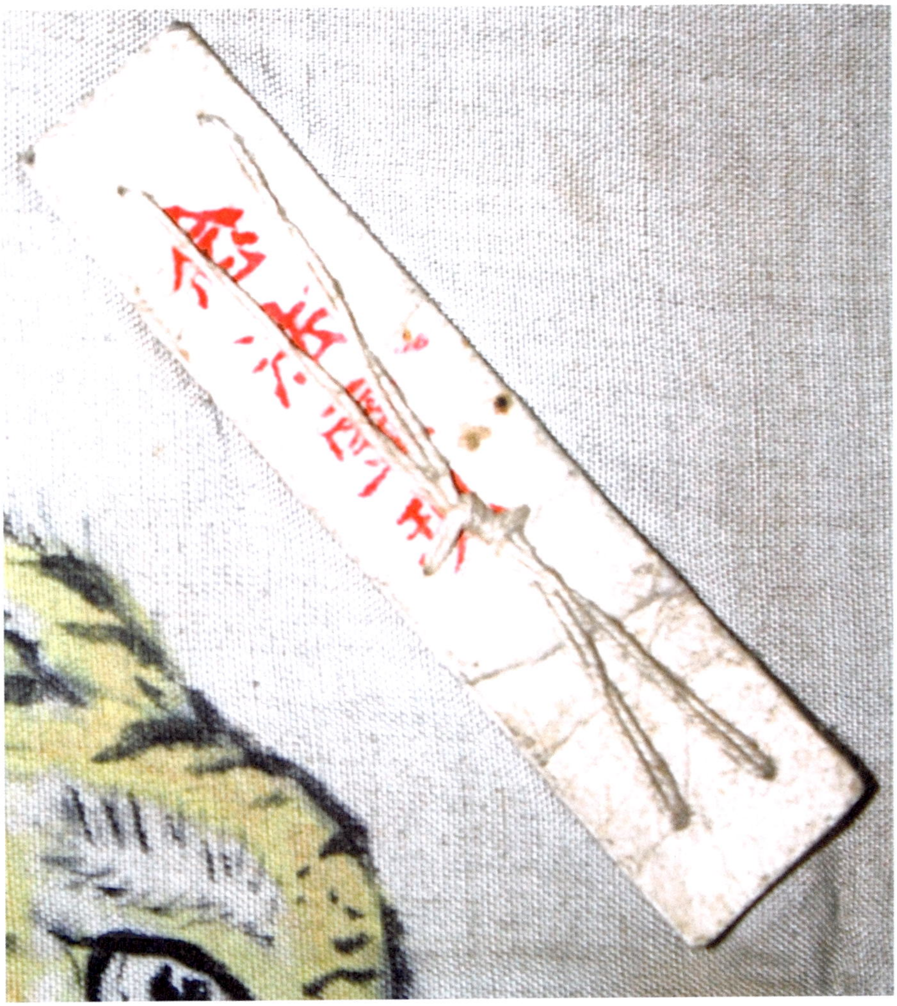

Tiger Good Luck Cloth/Amulet – 13.5" x 42.5" (34.29cm x 107.95cm). *Amulet courtesy of T.D.*

The nicely detailed tiger good luck amulet (bottom of page 165) was painted on un-hemmed, white cotton cloth. The tiger is large and well executed; the artist using pale shades of white, black, gray, red, orange and yellow ink or paint. It was presented as a gift to its owner and may have been used as a towel (*tenugui*) or perhaps a headband (*hachimaki*). Perspiration, as well as paint stains, are evident on the left-hand side and near the tiger's head.

The two small black *kanji* characters seen above the round, red seal near the tiger's back paws probably belong either to the person presenting the gift, or perhaps the artist. Unfortunately, the name is difficult to read. To the right of that, four larger vertical characters are for the slogan *Buun Chokyu*, or "May Your Military Fortunes Be Long Lasting". The longer vertical column of characters to the right, together with the single character on the far right, combine to say, "[Presented As A Gift To] Mr. Yamanaka Michisuke".

Tiger Good Luck Cloth/Amulet - 5.5" x 10.25" (13.97cm x 26.03cm). *Amulet courtesy of John Egger.*

This powerfully muscular tiger was painted upon a piece of cotton cloth in shades of black, white, gray and gray-brown ink. Artists often adopt a pseudonym when they sign their artwork, in much the same way that writers might select a pen name. Along with that, the characters that they use to write that name, as well as the stylized script encountered in their *inkan*, or name stamp, may prove difficult to read as well. In this case, both are very difficult to translate.

Section of a painting executed on a multi-sectioned Japanese folding screen. The overall work was titled, "Dragon and Tiger" and was created by one of the last *Kano* style artists, Hashimoto Gaho, in 1895 (the dragon was painted upon an adjacent panel). Exhibited at the Tokyo Metropolitan Art Museum in 1927, it was part of the *Meiji Taisho* Masterpiece Exhibition, sponsored by the Asahi Shimbun news organization.

Tiger Good Luck Cloth/Amulet - 8" x 13" (20.32cm x 33.02cm). *Amulet courtesy of John Egger.*

Image of a pouncing tiger painted on a remnant of cotton cloth using black and red ink. It contains no *kanji* writing or ink stamp of any kind. It was found among a number of other wartime era belongings owned by someone who served in the Japanese military during the war.

Tiger Good Luck Cloth/Amulet – 12" x 12" (30.48cm x 30.48cm). *Amulet courtesy of John Egger.*

Marvelous sitting tiger image painted upon cotton cloth, entirely in shades of black, gray and white. The vertical row of four *kanji* characters shown in the upper right-hand corner say *Kogun Muteki,* or "Unbeatable Imperial Army". The four ideograms in the corresponding upper left-hand corner say *Buun Chokyu,* or "May Your Military Fortunes Be Long Lasting". The probable artist's name and red seal in the lower left-hand corner are very difficult to read.

Tiger Good Luck Cloth/Amulet – 15" x 14" (38.1cm x 35.56cm). *Amulet courtesy of John Egger.*

Detailed painting of a tiger's head produced on cotton cloth using black, gray, red and white ink. Below the tiger's ear are three vertical rows of *kanji* characters. On the right-hand side, the five characters say *Ki Buun Chokyu,* or "I Pray Your Military Fortunes Are Long Lasting". The smaller vertical row of *kanji* in the center say *Tsuchinoe-Tora Ku-gatsu,* or "September 1938". The written name to the left is difficult to read, while the red seal below it is obscured.

Tiger Good Luck Cloth/Amulet - 12" x 16" (30.48cm x 40.64cm). *Amulet courtesy of Stephen Wallace.*

This colorfully detailed tiger good luck presentation cloth amulet was completely hand-painted on a piece of age-toned silk material. The artist blended black, white, gray, yellow, orange and brown paint to portray his attacking tiger image. The quality and detail of the painting points to a professional commission or a highly adept amateur.

The tiger measures over 10" (25.4cm) long, while the red sun, or *hinomaru,* is over 6" (15.24cm) in diameter. There are three red *inkan,* or seals, placed upon the flag. The smallest one measures 0.5" (1.27cm) high, while the largest measures about 0.75" (1.9cm) high. The round, red seal in the upper left-hand corner was hand-painted rather than a stamp. The translation of some of the *kanji* characters is as follows: "Dedicated To Masaharu Saito For His Second Enlistment In The Army!" There also appears to be a 1945 date, making this a very late-war presentation piece.

The End

"You might ask shockingly: 'A flag? What's that?
A stick with a rag on it?'
No sir, a flag is much more.
With a flag you lead men, for a flag, men live and die.
In fact, it is the only thing for which
they are ready to die in masses, if you train them for it.
Believe me, the politics of an entire group... can be manipulated
only through the imponderables that float in the air."

– *Theodor Herzl*

Glossary

Ainu – An ethnic people group indigenous to Hokkaido, Japan.

Amaterasu – Deity of the *Shinto* religion, as well as a part of the legendary Japanese pattern of formation myths. She is not only the goddess of the sun but of the entire universe as well. Her name comes from the word *Amateru* which means Shining in Heaven.

Amida Buddha – Also known as Amitabha, his name means Infinite Light. According to the scriptures of *Mahayana* Buddhism, he is known as a celestial Buddha. The Japanese refer to him as Amida Butsu.

Animism – A belief that attributes a soul to plants, inanimate objects and natural phenomena.

Anzaisho – Temporary lodging built to accommodate an imperial visit.

Arisaka – Japanese bolt action service rifle. The company first began manufacture around 1897 and stopped production with the end of World War Two in 1945. Arisaka rifles were accompanied by bayonets, made specifically to fit with each model of rifle.

Atelier – Workshop or studio. (Especially one used by an artist or designer).

A-Um – Sacred sound (Om) or spiritual icon in the Hindu religion. In Buddhism, it is recognized as a mantra.

Austronesia – The name given to a variety of people groups that inhabit parts of Southeast Asia, Oceania and East Africa, and speak languages that stem from the Austronesian family.

Baby Boom Generation – The generation of children born following the end of World War Two, typically considered to be from about 1945-1964.

Banzai - Live Forever, or Hail. (Commonly seen on Japanese good luck flags).

Battle of Minatogawa – A famous battle, also known as the Battle of Minato River, fought on July 4, 1336 between the forces of Emperor Go-Daigo, and those loyal to the Ashikaga clan. The Emperor's forces lost the battle, but it was also here that Kusunoki Masashige displayed his loyalty to the Go-Daigo, by offering up his life for his lord.

Battle of Shizugatake – A decisive battle that took place in May 1583 between the forces of Hideyoshi Toyotomi and Katsuie Shibata. Hideyoshi's army was victorious, with Katsuie committing suicide. A major outcome of the battle was the promotion of seven important generals (The Seven Spears of Shizugatake), including the tiger battling Conqueror of Korea, Kato Kiyomasa.

Beiei Gekimetsu - Destroy The United States And Great Britain! A common slogan seen placed upon Japanese good luck flags.

Bishamonten – Also known as Bishamon; he is a god of warriors who used a tiger as his personal messenger.

Bodhisattva – Sanskrit word for anyone that has generated *Bodhicitta*, which is a spontaneous wish and a compassionate mind in order to attain Buddha-hood for the welfare of all sentient beings.

Bonten – Known as the god Brahma in India.

Brahma – A four-faced creator god in Hinduism and the creator of the four *Vedas*, one coming from each mouth.

Bring-Backs – Term applied by many servicemen to their legally obtained war trophies. Items brought back generally required military paperwork that authorized service members to return home with their highly-prized souvenirs.

Buddha – Siddhartha Gautama, also known as Buddha. Buddhism was founded upon his teachings. He probably lived in Eastern India sometime between the 6th and 4th centuries B.C.

Buddha of Healing – Also known as the Medicine Buddha, Yakushi Nyorai, or Bhaisajyaguru. It is from this deity that the Yakushi-ji gets its name.

Buddhism – A philosophy or religion founded over 2,500 years ago in India by Siddhartha Guatama (also known as Buddha), that spread throughout much of Asia. There are numerous Buddhist sects, but teachings are based upon the Four Noble Truths. Buddhism was introduced to Japan around the year 550 A.D.

Bunraku – A traditional Japanese puppet play.

Bunten Exhibition (1907-1918) – The Japan Art Academy was founded in 1907 as the Fine Arts Reviewing Committee of the Ministry of Education. It was intended to establish standards of quality and to contribute venues for artists to display their works. Disagreement arose when artists complained that the judges were biased in selecting which artists could display at the show.

Bushi – A Japanese warrior, usually associated with a samurai.

Bushido – The way of the warrior. It contained an unwritten code of honor, initially followed by the Japanese samurai class, and included the fact that one must be willing to freely give up one's life for lord or emperor.

Buun Chokyu – May Your Military Fortunes Be Long Lasting. (The most prolific of slogans placed upon the Japanese good luck flag).

Bu Yu – Valor.

Chihaya Castle – Japanese castle constructed in 1332 by Kusunoki Masashige. The fortifications served as his primary defensive base of operations during campaigns in the service of his lord, Emperor Go-Daigo.

Chikamatsu Monzaemon – Japanese playwright who penned the famous puppet play *The Battles of Coxinga* in 1715.

Chikara - Power or Strength. (Normally represented by a single *kanji* character seen placed upon Japanese good luck flags).

Chougosonshi-ji – A Buddhist temple in Nara Prefecture, located half way up Mt. Shigi. The structure acts as the primary temple of Bishamonten and was built by Prince Regent Shotoku, in 594, on the spot where Bishamonten was said to have first appeared in Japan in the month, year and day of the tiger.

Chu – Loyalty.

Churetsu Mukyu – Absolute Loyalty And Never Any Difficulties.

Chuyu – Loyalty and Bravery. (Commonly penned upon Japanese good luck flags).

Confucianism – Classical Chinese philosophy, based upon the tenants of its founder Confucius. His teachings promoted a moral individual based on right interactions between individuals and social systems. A person's ancestors were given much importance in Confucian thought along with Classical studies. Confucius lived from 551-475 B.C.

Corner Tabs – Reinforced areas sewn into two of the four corners of a Japanese good luck flag during the manufacturing process. Corner tabs have been observed as being made from silk, paper, leather and other materials. They allowed the good luck flag to be hung up by strings without tearing the material.

Daibutsu – Another name for the Great Buddha of Kamakura or Amida Buddha.

Daimyo – Powerful Japanese feudal lords who remained in power until about the Meiji Period.

Dai Nippon Tei Koku Banzai – Long Live The Great Japanese Empire.

Dan – Determination.

Daruma – Buddhist monk who lived during the 5th or 6th century, also known as Bodhidharma. He is thought to have brought Chan Buddhism to China.

Deity of the Place – Tiger deity sent down to earth to rescue the Chinese Princess, Mia Shan.

Ebi – Shrimp. The sea creatures symbolize Longevity and the Renewal of Life.

Edo Period (1603-1868) – The period of time when Japan was under the rule of the Tokugawa shogunate. During this time there was great economic growth and a flourishing of the arts and culture. It was, however, a time of significant isolationism with the rest of Japan's neighbors and the outside world.

Ehime Prefecture – Prefecture located in northwestern Shikoku, Japan.

Eight Auspicious Symbols of Buddhism – When combined, these symbols are known as the *Ashtamangala*. In Buddhism, the Eight Auspicious Symbols of good fortune represent the offerings made by the gods to the Buddha upon his attaining enlightenment.

Emperor Jimmu – According to ancient lore, Jimmu was the first Emperor of Japan, ascending the throne around the year 660 B.C. Legend states that he is a descendant from the sun goddess Amaterasu, through Ninigi, her grandson.

Emperor Keiko – According to legend, the 12th Emperor of Japan, ruling sometime between 71-130 AD.

Emperor Meiji – Japanese Emperor from 1868-1912.

Escheatment – The process of land confiscation from the samurai during the Tokugawa era.

Essence – Philosophically, essence is the characteristic or characteristics that make a being or thing what it essentially is. If those qualities are missing, then so too is the being's existence.

Fibroin – This protein is the basic structural component of de-gummed silk fiber.

First Sino-Japanese War (1894-1895) – The war between Japan and China to see who would take control of Korea.

Fly Side – The side of a flag that blows in the wind, opposite the hoist side which is usually tied to a flag pole. It is generally reinforced with extra hem stitching to help prevent fraying.

Fragrant Island – The location where The Deity of the Place, in the form of the tiger, transported Mia Shan so that she could attain enlightenment.

Fuji – Protector (as in an emergency).

Fukushima Prefecture – Prefecture in the Tohuko region of Japan, on the island of Honshu.

Fukutora – The world's largest *papier-mâché* tiger, located at the Chougosonshi-ji Temple.

Gambare – Do Your Best.

Ganges Delta – A river delta in the Bengal region of Southern Asia.

Ganjosai Kunihiro (active between 1815-1843) – Japanese woodblock print artist.

Genko War Period (1331-1333) – War marking the end of the Kamakura shogunate which brought about the Kemmu Restoration.

General Kanki – Chinese General married to Watonai's half sister Kinshojo.

Giyu Hoko – Serving With Justice And Bravery.

Gobun – Crushed shell paint used to cover Japanese tiger *hariko* overall uniform white before the final colors are applied.

Go-Daigo – 96th Emperor of Japan, according to the traditional order of succession. Born in 1288 and died in 1339 from an unknown disease.

Go Funto Wo Inoru – Praying For A Strong Fight.

Goseikun – Precepts or the True Way (as taught by the Buddhist sect Nenpo Shinkyo).

Go-Sen – Five-sen, a denomination of coin sometimes found sewn onto a good luck flag, or *senninbari,* as an amulet of good fortune. The number four is unlucky in Japanese culture, and the five-sen coin represents going over or beyond four, as well as to move past the line of death.

Great East Asia War – In Japanese, it describes the war that initially broke out during the formation of the Greater East Asia Co-Prosperity Sphere by the Japanese Empire. It involved nations within Northeast and Southeast Asia as well as surrounding areas, and it spread beyond. The War lasted from the mid-1930s until 1945.

Guadalcanal – Principal island of the Solomon Islands. During World War Two, an engagement was fought here between Japanese and United States forces from August 7, 1942 until February 9, 1943.

Gunka – Patriotic song.

Gyoko – Emperor's Visit.

Hachimaki – Headband or bandanna worn around the head by the Japanese during times of strenuous effort or physical activity. Some *hachimaki* were decorated with designs or had slogans painted or printed upon them.

Han Dynasty (206 BC-220 AD) – The second Imperial dynasty of China. Spanning over four centuries, it is considered a golden age of Chinese history, and economic vitality.

Hanko – Stamps or seals believed by some to have a magical power that may bring about success. Personal stamps, known as *inkan* carry as much importance as do personal signatures. They are used in many aspects of daily life, especially to endorse personal documents. They are often made from bone, wood, ivory or other material.

Hariko – Japanese folk dolls.

Hashimoto Gaho (1835-1908) – Japanese painter who was one of the last to paint in the Kano School style. He created a number of works of art that utilized combinations of color and gold or monochrome black ink. Many of his paintings portrayed traditional subjects, while at the same time he incorporated Western art methods. Brush strokes, various detailing and perspective are all evident in his work.

Hashimoto Keibu – A minor samurai whose descendants first began to make miharu ningyo (folk dolls).

Hata – Flag.

Heian Period (784-1185) – Period of the last division of Japanese classical history. Named after Heian-kyo, today's Kyoto.

Hichesei Metteki – Seven Lives To Decimate The Enemy.

Hinomaru – The Japanese national flag comprising a white field with a round, red sun near the center. Early folklore says that the flag was first presented as a standard to the Emperor by the Buddhist priest Nichiren at the time of the Mongol invasions in the thirteenth century.

Hissho – Certain Victory!

Hissho No Shin Ne – Faith In Certain Victory.

Hoist Side – The side of a flag that is usually tied to a flag pole. On most Japanese good luck flags, the hoist side contains two reinforced corner tabs made from various materials, and two tie strings normally made from silk or cotton.

Hokan – Crown. The Chinese Princess Mia Shan wore a *hokan*.

Honshu – The largest and most populated island of Japan.

Hosso Buddhism – Consciousness-Only Buddhist sect which came to be called Hosso in Japan.

Ideograms – Graphic symbols that represent an idea, rather than a group of letters arranged according to the phonemes (representations of sound) of a spoken language, as is done in alphabetic languages.

Igorot – Austronesian group of people that collectively live in the mountain region of Luzon, Philippine Islands.

Indra – One of the twelve guardian deities in Buddhism who protect Buddhist teachings and believers. He is known as Taishakuten.

Inkan – In Japan, seals are generally referred to as *hanko* or *inkan*. Special seals may have a person's name carved or engraved upon them. These seals may carry the same weight as a personal signature.

Issei Banteki Iatsu – Dominate The Enemies With One Shout!

Iwo Jima – An island of the Japanese Volcano Island chain, south of the Ogasawara Islands. Together they form the Bonin Islands. During World War Two, Japanese and United States military forces fought a battle here from February 14 until March 26, 1945.

Jinchu Hokoku – To Make Sacrifices For The Emperor/Country.

Jinja – The name given to *Shinto* shrines.

John Wayne (1907-1979) – An American actor who starred in over 142 films, many based upon U.S. military themes.

Kami – The spirits within objects, as described in the *Shinto* belief system.

Kamikaze – Divine Wind. *Kamikaze* were also members of the Special Attack Corps, or *tokkotai,* trained during World War Two to sacrifice their lives by flying a plane, piloting a torpedo, etc. into an enemy ship.

Kamon – A crest used in Japan to indicate family lineage, etc.

Kanji – A form of Japanese alphabet. As a writing system, *kanji* is based upon ideographic symbols (pictograms) originally designed by the Chinese.

Kano School – A famous Japanese school of painting, dominant from the late 15th century through 1868.

Kansai Region – Region of the south-central portion of Japan on the main island of Honshu.

Kanto Region – Region of the island of Honshu that includes Tokyo and six prefectures.

Kapok Tree – Scientific name *Ceiba pentranda*. A tree whch provides a buoyant material, much lighter than cork, used in pilot's float flags and life preservers.

Karma – Sanskrit word that refers to the principal of cause and effect whereby one's intent and actions influence the future of that individual.

Kawarasaki Gonjuro 1 (1838-1903) – One of the most successful and famous *kabuki* actors of the Meiji period (1868-1912). Near the end of his life, he was given the honorary name, Ichikawa Danjuro IX.

Keian Uprising – Failed coup attempt in 1651 against the Tokugawa shogunate by Yui Shosetsu, Marubashi Chuya and a number of *ronin*. Named after the Keian era in which it took place, the uprising was emblematic of the problem the government faced with disenfranchised samurai.

Kemmu Restoration – Failed attempt by Emperor Go-Daigo to bring the Imperial House back into power. Its success would have restored a civilian government after almost 150 years of military rule. Led instead to the rise of the Ashikaga shogunate.

Kenyu – Strictly Courageous.

Ki Buun Chokyu – I Pray Your Military Fortunes Are Long Lasting.

Kikugawa Eiji – The father of well known woodblock print artist, Kikukawa Eizan. Eiji was a minor Kano School artist who taught his son early in his career.

Kikukawa Eizan (1787-1867) – Designer of *Ukiyo-e* style woodblock prints. He initially studied under his father, Kano School artist Kikugawa Eiji. His style was reminiscent of Utamaro, exhibiting lightness and grace.

Kikusui no Mon – Emblem from the family crest of samurai warrior Masashige Kusonoki. The badge depicts a half chrysanthemum floating above a flowing river.

Kinshojo – From the play *The Battles of Coxinga*. Kinshojo was the half sister of the Japanese warrior Watonai-Coxinga; she was married to the Chinese General Kanki.

Kitagawa Utamaro (1753-1806) – One of the most highly regarded designers of *Ukiyo-e* woodblock prints and paintings, best known for his large head prints of women, primarily of the 1790s. Utamaro's work even reached Europe where his influence may be seen in the works of the Impressionists.

Ki Tora Shin – The Wooden Tiger God.

Kiyoshi Nobutoki – Composer who took some of the lyrics from the poem written by Otomo no Yamamochi and used them in his 1937 patriotic song *Umi Yukaba*. That song came to be known as Japan's second national anthem.

Ki Zensho – Prayers For Complete Victory.

Kobe – Sixth largest city in Japan, located on the south side of the island of Honshu. On April 18, 1942 in a daring attack, Kobe and Tokyo were both bombed by the Doolittle Raiders.

Kobo-Daishi (774-835) – Japanese Buddhist monk who founded the Shingon school of Buddhism.

Kongaiji – Primary temple of the Nenpo Shinkyo Buddhist sect in Osaka, Japan.

Kongorikishi – The two guardian figures assigned to protect the Buddha during his travels, known by their names Misshaku Kongo and Naraen Kongo. They are also emblematic of the Buddhist mantra *A-um* or *Om*.

Konohanasakuyahime – Seen in Japanese mythology, she is also known as Konohana. As a deity, she carries the title of cherry blossom princess and is the daughter of the mountain god, Oyamatsumi.

Koriyama Naval Air Group – Japanese unit formed on February 1, 1944 but dissolved about a year later. March 15, 1945 it was redesignated 1st Koriyama Naval Air Group.

Koshin no Tora – Yellow Bodied Tiger.

Kougeki Seishin – Attack Spirit. Japan attributed its victory in the Russo-Japanese War (1905-1905) to the aggressive spirit of its soldiers. The characteristics that came to dominate that phrase were inculcated within that and Japan's future military personnel.

Koyama Chikusai – Meiji era artist, active between 1868-1912.

Kumbh Mela – A mass Hindu pilgrimage where those in attendance bathe in a sacred or holy river.

Kun – Used to address men who are younger or the same age as the speaker. Often a term of affection given by an older relative to a younger one. Also applies to one of equal or lower status.

Kusunoki Masashige (1294-1336) – 14th century samurai warrior who led troops that fought for Emperor Go-Daigo in the Genko War.

Kyokujitsu-ki – Japanese battle flag.

Kyoto – Ancient capital of Japan and early home of the Emperor, founded in 794 A.D.

Kyuteki Gekimetsu – Decimate The Enemy.

Lotus Sutra – One of the most influential *sutras,* forming the basis on which the Nichiren and other schools of Buddhist thought were established. Many believe it encapsulates the Buddha's last teaching for salvation.

Makoto – Faithful.

Mantra – Sacred sound that, when repeated, is believed to have the ability to alter a person's state of consciousness. A mantra may or may not have an actual meaning.

Man' Yoshu – Oldest existing collection of Japanese poetry composed after 759 A.D. during the Nara period (710-794).

Marubashi Chuya – Famous samurai who co-conspired with Yui Shosetsu to overthrow the Tokugawa shogunate in 1651.

Material Culture – Referring to physical objects, resources and spaces that people use to define their culture.

Matsunoki – Pine tree.

Mayoke Tora – Tiger Amulet.

Meiji Period – Time frame from 1868-1912. The Emperor at this time was known as the Meiji Emperor, and his rule was known as the Enlightened Period.

Mia Shan – Also known as Guanshiyin in China, she is an East Asian Bodhisattva, primarily venerated by Mahayana Buddhists and often called The Goddess of Mercy.

Miharu Ningyo – They take the forms of protective dolls.

Mikawa Bay – Bay located to the south of Aichi Prefecture, Japan.

Million Dollar Movies – Television movie series begun in New York City in 1955 and ended in 1966. Many of its films were broadcast twice daily and were shown for an entire week.

Mimibukuro – Essay that, early on, more fully introduced the idea of Samuhara to the Japanese public.

Minamoto no Yoshitsune (1159-1189) – Military commander of the Minamoto clan in the late Heian and Kamakura periods. Remembered as one of the greatest and most popular warriors of his era, and one of the most famous *samurai* warriors in the history of Japan. He was forced to commit *seppuku* after being betrayed by the son of a trusted friend.

Ming Dynasty – Ruling dynasty of China between (1368-1644).

Misshaku Kongo – The open-mouthed guardian of the Buddha. The open mouth signifies the first part of he sound *O-* for the mantra *O-Um.*

Mo Ko Zokugo – Continuous Ferocious Attack!

Mount Fuji – The highest mountain in Japan, located on the island of Honshu. For centuries, Fuji-san has been of as a sacred mountain.

Mount Shigi – Mountain in the Ikoma District of Nara, Japan. While at Mount Shigi, Prince Shotoku of the Soga clan prayed to the God of Warriors, Bishamonten, for victory in his religious battle with the Mononobe family. It was said that Bishamonten aided the Prince's forces, and following their win, the Prince built a shrine here in his honor.

Mount Torafusa – Mountain known as A Tiger Leaning On Its Side.

Muromachi Period (1336-1573) – Period marked by the rule of the Muromachi shogunate. Officially recognized in 1338, the period lasted until 1573 when its 15th and final shogun was exiled from the capital of Kyoto.

Musashi – Named after the former Japanese Province, *Musashi* was the second of two super battleships constructed for the Japanese navy during World War Two. Along with her sister ship, *Yamato*, she was one of the largest and most heavily armed battleships ever built.

Musha-e – Warrior Pictures. Pertaining to *Ukiyo-e* art that captures the image of the traditional warrior.

Nagasaki – The capital and largest city of Nagasaki prefecture on the island of Kyushu. Second city (after Hiroshima) to be devastated by an atomic bomb during World War Two.

Nagoya – The largest city in the Chuba region on the island of Honshu.

Nakaime O Ikiru – To Live In The Moment.

Nammyouhorengekyo – Central mantra chanted within all forms of Nichiren and Tendai Buddhism.

Nara – City in southern Honshu, founded n 706 A.D. and its capital from 710-784 AD. Numerous cultural and religious advances of the day came from here.

Nemari Tora – Crouching Tiger. Nemaru comes from the Yamagata dialect and means "crouching". Related images show tigers with their forelegs in a position as to jump or pounce right away.

Nihon Shoki – The Chronicles of Japan. This ancient text is the second oldest book that details classical Japanese history and lore.

Nihonto – Japanese sword.

Ninigi-no-Mikoto – In Japanese mythology he is known as the grandson of Amaterasu, the sun goddess.

Nirvana – In Buddhism, nirvana represents that state whereby a person is finally liberated from the endless cycle of death and rebirth.

Nishiki-e – Japanese multi-colored woodblock printing method used primarily in *Ukiyo-e* art. In this format, each color of ink is applied using a separately carved wood block.

Nitten Exhibition – In 1947, the Imperial Art Academy transitioned to the Japan Art Academy. Its annual art shows were then renamed the Japan Arts Exhibition, or Nitten, for short.

Nobori – Banner.

Ofuda – Honorable Plaque. Traditional good luck item, written on *washi* paper, consisting of a prayer sanctified by a priest and used to ward off evil spirits and attract good luck. Some *ofuda* may be folded inside pieces of cloth or stitched to the outside of the material. Others are carried in special charm pouches.

Ogata Gekko (1859-1920) – Self-taught painter and designer of *Ukiyo-e* prints. Stylistically, his work was inspired by the artist Hokusai. Many of his well-known works covered subjects from the First Sino-Japanese War.

Ogatama – Inviting Soul.

Okubi-e – Large head style of art popularized by a number of woodblock print artists.

Omamori – Popular Japanese good luck charms or amulets normally sold at *Shinto* shrines or Buddhist temples. They are said to provide good luck to those who purchase them. They are often dedicated to Buddhist or *Shinto* deities.

Orientalist – Someone from the West who studies the language, culture, history or customs of countries in Eastern Asia.

Osaka – City in the Kansai Region. Third largest city in Japan (by population) after Tokyo and Yokohama.

Osaka Art College – College founded by the famous Japanese artist Yano Kyoson.

Ota Castle – Focal point for the forces of Toyotomi Hideyoshi during the siege of Negoro-ji/Ota Castle in 1585. Hideyoshi's forces attacked and burned the castle complex following the initial siege and departure of the *saiga ikki* warrior monks from Negoro-ji. After Ota's surrender, Hideyoshi ordered all remaining *samurai* killed, but allowed the peasant defenders to lay down arms and return to the fields of their lords.

Otasuke Kannon – Divine Help From The God Of Compassion.

Otomo no Yakamochi – Famous Japanese *waka* poet and statesman, and one of the compilers of the *Man' Yoshu* (Japanese book of poetry).

Oyamatsumi – The mountain god. His daughter, Konohana, is known as the cherry blossom princess.

Papier-mâché – Composite material of paper pulp, textiles and water/glue. *Papier-mâché* dolls are created by pressing the composite around a molded wood shape and allowing it to dry. Afterward, the pieces may be glued together and painted.

Pearl Harbor – Lagoon harbor on the island of Oahu, Hawaii. The United States Navy's Pacific Fleet is located there. On December 7, 1941 Japanese naval forces launched a sneak attack upon the United States military based there. 2,403 servicemen and civilians were killed in the attack.

pH Scale – Numeric scaled used to describe acids and bases within aqueous solutions.

Pilot's Float Flag – Japanese national flag made from silk and lined around all four edges with a buoyant material known as kapok. The flag was tied by string to a collapsible wooden pole and carried in a pouch on the life preserver. In case of an emergency, they could be waved in the air or floated upon the water while awaiting rescue.

Prince Shotoku Taishi (574-622) – Semi-legendary ruler of Japan. Related to the Soga clan, he popularized Buddhism (at the time seen as a liberal belief system) in Japan. At the same time, he maintained his respect for the earlier *Shinto* system of beliefs.

Qin Dynasty (221-622) – The first Imperial Chinese dynasty. The Qin wanted to create a system of structured political power that fed into a large military, in conjunction with a stable economy.

Riki – Strength.

Ronin – A samurai warrior without a lord or master during Japan's feudal period (1185-1868). This normally occurred when the lord was killed or died without an heir.

Russo-Japanese War (1904-1905) – War between Russia and Japan over both country's ambitions in Manchuria and Korea.

Sacred Treasures – Japanese Imperial regalia also known as the "Three Sacred Treasures", comprising the Sword of Kusanagi, the Mirror of Yata no Kagami, and the Jewel of Yasakani no Magatama. They represent the three virtues of valor, wisdom and benevolence.

Saga Prefecture – Prefecture in the northwest portion of the island of Kyushu.

Saiga Ikki – Buddhist warrior monks, led by Saika Magoichi, based in Ota, in Kii Province. Many of them were killed by the forces of Toyotomi Hideyoshi during the Siege of Negoro-ji/Ota Castle in 1585

Saika Magoichi – Leader of the Saiga Ikki warrior Buddhist monks defeated at the Siege of Negoro-ji/Ota Castle.

Sainan Yoke No Tora Yakushi – Tiger Yakushi To Prevent Misfortune and Disaster.

Saipan – The largest of the Northern Mariana Islands. During World War Two, Japanese and United States military forces fought a battle that lasted for three weeks, beginning on June 15, 1944.

Sakoku – The period of isolation that existed in Japan during the Tokugawa shogunate. It began c.1633 and ended c.1853 with the arrival of "The Black Ships" under the command of Matthew Perry.

Sakuteiki – The oldest Japanese book on gardening (and possibly the oldest in the world), written sometime in the mid-to-late 11th century.

Samuhara – *Shinto* mantra that asks the divine spirits to blanket the entire person with protection, including mind, body and spirit. It specifically acts to protect the wearer from harm caused by bullets in battle.

Samuhara Jinja – *Shinto* shrine established by Tanaka Tomisaburo to promote the Samuhara good luck amulet.

Samurai – Military nobility and officer caste of medieval and early modern Japan.

Second Sino-Japanese War (1937-1945) – Conflict initially between China and Japan over control of the Chinese mainland. When the Japanese attacked Pearl Harbor in 1941, the Second Sino-Japanese War expanded into World War Two.

Sei Hei – Sacred Soldier.

Sei Ken – Sacred Progress.

Sei Sen – Sacred War.

Sei Shi – Sacred Death.

Sei Sho – Sacred Victory.

Seishin Itto – Unwavering Spirit.

Sen – Japanese monetary unit equal to one-hundredth of a yen.

Sendai-han – Domain established near the end of the 16th century (during the Edo period) and ruled by the Date clan, whose land and family castle covered most of present day Miyagi Prefecture.

Sengoku Period (1467-1603) – Period in Japanese history marked by social turmoil, political in-fighting and war.

Senninbari – Good luck amulets of protection made by women and given to Japanese soldiers going off-to-war. *Senninbari* are most commonly seen in the form of belts, known as one-thousand stitch belts. Translated into English, *senninbari* means "one thousand person stitches". *Senninbari* took various forms including belts, vests, caps, scarves, handkerchiefs, flags, etc.

Sensei – Teacher.

Seppuku – Cutting the stomach/belly. Japanese ritual suicide.

Sericin – Protein, also known as silk gum. In the process of making silk material, sericin is removed.

Seven Great Temples Of Nara – Also known as Nanto Shichi Daiji. The common historical name refers to the influential "Seven Buddhist temples", home of the Hosso School of Japanese Buddhism in Nara/Nanto. The Yakushi-ji is one of the most famous in the group.

Shi – Japanese for the number four, it is also the word for Death.

Shido Gi Wo Oite Bakusai – Enormous Warrior Spirit Through Righteousness.

Shingon – One of the most significant schools of Japanese Buddhism. It was originally spread from India to China, and then Japan, by traveling monks. *Shingon* is the Japanese reading of the Chinese that means "True Words" or "Mantra" in Sanskrit.

Shinpu Tokubetsu Kogeki Tai – Divine Wind Special Attack Units.

Shintai Kenzen – Good Health.

Shinto – Early, dominant religion of Japan prior to the introduction of Buddhism. *Shinto* has no formalized dogma but emphasizes the need to be loyal to one's ancestors. *Shinto* became dominant during World War Two and has a pantheon of gods. Ameterasu (the sun god), is the chief god from which the Japanese Imperial Family claim descent.

Shioto Town – A town in the Fujitsu District, Saga Prefecture in Japan. In 2006 it merged with Ureshino Town to form Ureshino City.

Shisei Hoko – Serving With Utmost Sincerity.

Shizuoka Prefecture – Japanese prefecture in the Chuba region, on the island of Honshu.

Shogun – The military ruler of Japan from approximately 1185 until 1868.

Showa Period (1926-1989) – Emperor Hirohito was known as the Showa Emperor. *Showa* means Enlightened Peace.

Shunsen Natori (1886-1960) – Considered by many to be the last master in the art of painting *kabuki,* or actor pictures. Nearly all of his portraits were done in the *okubi-e,* or large head style.

Shussei Nobori – Off-to-war banners presented to servicemen leaving home for military training or battle overseas.

Sichisei Hokoku – Seven Lives In Devotion To The Nation/Country.

Siege Of Negoro-ji – Led by Toyotomi Hideyoshi in 1585. Prior to the siege, Hideyoshi's forces had attacked a number of nearby fortifications manned by warrior monks. Next, he attacked the Negoro-ji complex and laid siege to it. Many of the temple's monks escaped and fled to the nearby Ota Castle. Hideyoshi's men set fire to the castle complex and killed its defenders as they ran from the flames.

Sinologist – A Western academic who studies China primarily through Chinese language, literature, Chinese culture and history.

Soga Clan – One of the most powerful clans of the Asuka Period (approximately 538-710). It played a major role in the spread of Buddhism throughout Japan.

Soshotai – Also known as cursive script, *soshotai* is utilized primarily as a shorthand method of writing as well as a popular calligraphic style. For those familiar with it, it is an easy style to read, however, it is used less often today in Japan, and is becoming somewhat obsolete.

Sugiyama Toshisuke (1886-1946) – Japanese Vice-Admiral and Commander of the Navy Technical Department during World War Two. During his tenure, he oversaw funding for, and was instrumental in having built, the super battleships *Yamato* and *Musashi.*

Sumitomo Steel – The company first began modern operations in 1897 when Sumitomo Copper Works was opened in Osaka. Soon thereafter (1901), the company began steel manufacturing and was a major producer of arms for the Japanese military during World War Two. It merged with Nippon Steel in 2012.

Sunpu – Present day Shizuoka City, located in Shizuoka Prefecture on the island of Honshu.

Sutra – The basic scriptures considered to be the oral records of the teachings of Gautama Buddha.

Taishakuten – Known as the god Indra in India.

Taisho Period (1912-1926) – Corresponding to the reign of the Japanese EmperorTaisho. *Taisho* means Period Of Great Righteousness.

Take Ni Tora – Tiger In Bamboo.

Tanaka Tomisaburo (1868-1967) – Former Japanese soldier who popularized the Samuhara good luck amulet during World War Two. The amulet/mantra was believed to protect its owner from bullets in battle.

Tarawa – An atoll in the Central Pacific Ocean. A battle between the military forces of Japan and the United States occurred here on November 20, 1943.

Teiten National Exhibition – Name given to the premier Japanese national arts exhibition in 1919.

Tekikoku Kofuku – Enemy Countries Shall Surrender.

Tensho Script – A form of calligraphy script from China. The fanciful *tensho* style is no longer favored as an everyday script in China. In Japan, *tensho* continues to be used primarily as a seal script.

Tenugui – A thin Japanese hand towel made from cotton.

The Chronicles Of Japan – Also known as the *Nihon Shoki.* The second oldest book that details classical Japanese history and lore.

The Fighting Seabees – World War Two era movie (1944) starring John Wayne and Susan Hayward. Based upon a fictionalized account that led to the founding of the United States Navy Seabees.

The Great Buddha of Kamakura – Statue located in Kamakura. Cast in 1252, it is the second largest Buddha statue in Japan.

The Seven Spears of Shizugatake – Seven samurai warrior bodyguards for Toyotomi Hideyoshi during the Battle of Shizugatake in 1583, instrumental in helping him to defeat the forces under his rival, Katsuie Shibata. As a reward, some were promoted to Daimyo. The title was later bestowed upon them within popular society.

The Three Great Ronin – Besides Yui Shosetsu, the other two famous ronin were Kumazawa Banzan and Yamago Soko.

The Twelve Divine Generals – The protector deities for Bhaisajyaguru, also known as the Buddha of Healing or Medicine Buddha.

Tie Strings – Refers to small white strings made from silk or cotton found attached at each of the corner tabs of a good luck flag.

Todaiji Temple – Buddhist temple complex in the city of Nara, Japan. Building began in 728 and it became known as one of the Seven Great Temples. It houses the largest statue of Buddha in Japan and today is the center of the Kegon School of Buddhism.

Tokkotai – Transcending Life And Death. Tokkotai were members of the Special Attack Forces, or *Kamikaze.*

Tokugawa Iemitsu (1604-1651) – The third Shogun of the Tokugawa ruling family. He was known to have persecuted Christians and expelled Europeans from Japan. Following this action, he had Japan's borders closed, a decision that remained in effect for about two hundred years.

Tokugawa Ieyasu (1543-1616) – The first Shogun and founder of the Tokugawa shogunate of Japan. They ruled Japan from about 1600 until 1868.

Tokugawa Shogunate (1600-1868) – The last Japanese feudal military government. The Shogun was the head of the government.

Tokyo – Capital city of Japan and one of its 47 prefectures. Tokyo is officially known and governed as a "metropolitan prefecture". It was originally a small fishing village known as Edo.

Tony The Tiger – Copyrighted advertising cartoon character and mascot for Kellogg's Frosted Flakes breakfast cereal, first appearing in 1952.

Tora – Tiger.

Toryu – The Climbing Dragon.

Toyohara Kunichika (1835-1899) – Well known Japanese woodblock print artist who, in his early years, studied under the famous print maker Utagawa Kunisada. *Kabuki* actors and beautiful women lent themselves as subjects to much of his art.

Toyokawashi City – City in the eastern portion of Aichi Prefecture. In 1939, the Toyokawa Naval Arsenal was established there.

Toyokawashi Naval Arsenal – Enormous arsenal built by the Japanese military in 1939, following talks that began in 1937. It was a major producer of aviation ordnance, light arms and ammunition for the Japanese Navy during World War Two. It was bombed by Allied forces following the atomic bombing of Hiroshima.

Toyotomi Hidenaga (1540-1591) – Half brother of Toyotomi Hideyoshi, and one of the most powerful warlords of the Sengoku Period.

Toyotomi Hideyoshi (1537-1598) – A *daimyo,* samurai warrior and politician during the Sengoku Period of Japan. Hideyoshi left behind numerous cultural legacies, including the restriction that only members of the samurai class were allowed to bear arms, as well as the restoration and rebuilding of many temples, etc. He was also prominent in the history of Christianity in Japan, following his order to execute, by crucifixion, 26 Christians.

Triptych – A work of art divided into three sections, or three carved panels.

Tsuka – The handle of a Japanese samurai sword.

Type 90 Steel Helmet – The standard helmet used by Japanese Navy and Army combat forces during World War Two.

Uchite Yamamu – To Attack And Decimate.

Uchu No Tora – Tiger In Rain. Symbolizes the relationship between the tiger and the dragon or *Yin-Yang* interplay.

Ukiyo-e – Pictures of the floating world. *Ukiyo-e* took the form of woodblock prints and paintings that were produced when blocks of wood with raised images were inked and the image was pressed onto paper. Pictures were initially created in black and white, with many being used to illustrate books. Later, colorful paints were added in order to produce bold and beautiful likenesses of various subjects, including *kabuki* actors, travel scenes, landscapes, animals, plants and erotic themed art. This style of art was popularized in the 17th through 19th centuries and is highly collected today.

Umi Yukaba – Popular patriotic song, written in 1937 and known as Japan's second national anthem.

Utagawa Kunisada (1786-1865) – Also known as Utagawa Toyokuni lll, he was the most popular and commercially successful artist of Japanese woodblock prints of the 19th century. Along with Hiroshige and Kuniyoshi, Kunisada is acknowledged as being one of the three greatest artists of the Edo period.

Utagawa Kuniyoshi (1798-1861) – One of the last great Japanese *ukiyo-o* masters, whose subjects included mythical animals, cats, *kabuki* actors and others.

Utagawa Toyonobu (1859-1886) – *Ukiyo-e* artist and print maker who lived in Tokyo.

Utagawa Yoshikazu (active 1848-1870) – *Ukiyo-e* artist who was active from the end of the Edo period to the first half of the Meiji period.

Utagawa Yoshitsuya (1822-1866) – Designer of Japanese woodblock prints, best known for his pictures of samurai warriors. Early in his career, he studied under Utagawa Kuniyoshi.

Waka – Genre of Japanese poetry.

Wakayama Castle – Famous castle at the mouth of the Kii River. Formerly called Ota Castle, it was the home of Saiga Ikki (warrior monks) who were killed by the forces of Toyotomi Hideyoshi at the Siege of Negoro-ji in 1585.

Wakayama Prefecture – Prefecture located on the Kii Peninsula in the Kansai region on the island of Honshu. The Sasutahiko Shinto Shrine is located here.

Washi – Traditional Japanese paper. The word comes from *wa-* meaning Japanese and *-shi* meaning paper.

Watanabe Nobukazu (1874-1944) – Student of Chikanobu Toyohara during the late Meiji period. His artwork was in the traditional woodblock print style. His subjects included various historical themes from the imperial period as well as Sino-Japanese, and Russo-Japanese Wars.

Yakushi-ji – One of the oldest and most famous imperial Buddhist temples in Japan, named after Yakushi Nyorai, or Medicine Buddha, one of the first Buddhist deities to arrive in Japan from China, in 680 A.D. Once part of the Seven Great Temples of Nanto in Nara. The Hosso School of Buddhism is headquartered on this site.

Yamato – One of the two Japanese super battleships built during World War Two. Along with her sister ship the *Musashi,* she was one of the largest and most heavily armed ships ever built.

Yamato Damashii – Phrase often used in World War Two by the Imperial Japanese armed forces to mean Japanese Fighting Spirit, intended to motivate the troops. The Yamato spirit has its antecedents in ancient Japanese history.

Yano Kyoson (1890-1965) – Famous Japanese artist born in Ehime prefecture on the island of Shikoku. During his career, he exhibited in all three Japanese art exhibitions: Bunten, Teiten and Nitten. He went on to establish the Osaka Art College.

Yokaren – The elite Naval Preparatory Aviation Training Program. Applications for the Yokaren were much in demand especially following the success of the Japanese at Pearl Harbor. Later in the war, some *tokkotai* pilots trained at the Yokaren.

Yokohama – Second largest city in Japan (by population) after Tokyo. It lies in the Kanto Region on Tokyo Bay.

Yokomuki Tora – Tiger Looking To The Side.

Yokosuka Naval Base – Japanese Naval facility, and the first of four administrative districts within operational control of the pre-war Japanese Navy. Later there would be others. Near the end of World War Two, the facility was bombed by the Allied air forces, but was surrendered nearly intact.

Yosegaki – To Write Sideways. The Japanese good luck signed flag, or *hinomaru yosegaki,* was commonly made from a Japanese national flag with inked names and slogans written on the white field in a generally sideways pattern.

Yoshida Toshi (1911-1995) – Print artist who spent his early career sketching animals. His father, the artist Yoshida Hiroshi, was a landscape painter and printmaker who insisted that his son follow similarly in his footsteps. It wasn't until about 1970 that Toshi returned to his first love, animal prints.

Yu Huang – The Supreme Heavenly Deity or First God.

Yui Shosetsu (1605-1651) – Military leader, swordsman and strategist who led an unsuccessful revolt against the Tokugawa shogunate in 1651 that came to be known as the Keian Uprising. He was recognized as one of the Three Great *Ronin* after his death.

Yusen Funto – Fight The Battle With Courage.

Zen – School of Mahayana Buddhism that originated in China during the *Tang* dynasty as Chan Buddhism. Zen was influenced strongly by Taoism. It is very popular today in Japan and worldwide.

Bibliography

Andrade, Tonio. *How Taiwan Became Chinese: Dutch, Spanish and Han Colonization in the Seventeenth Century.* Columbia University Press, New York, 2008.

Ashton, W. G., Terrance Barrow. *Nihongi: Chronicles of Japan from the Earliest Ties to AD 697.* Routledge Publishing Co., Oxford, 2010.

Baird, Merilee. *Symbols of Japan.* Rizzoli International Publications, New York, 2001.

Bartlett, John. *Bartlett's Familiar Quotations.* Little, Brown and Co., New York, 2002.

Bentley, John R. *A Descriptive Grammar of Early Old Japanese Prose.* Brill Publishers, Inc., Boston, 2001.

Berthier, Francis. T*he Japanese Dry Landscape Garden: Reading Zen in the Rocks.* University of Chicago Press, Chicago, 2008.

Biedermann, Hans. *Dictionary of Symbolism.* Penguin Books, New York, 1994.

Black, Candice (editor). *Tiger: 100 Representations in Classic Japanese Art.* Solar Books, Chicago, 2012.

Bortner, Michael A. *Imperial Japanese Good Luck Flags and One-Thousand Stitch Belts.* Schiffer Publishing, Atglen, 2008.

Bottomley, I. and A. Hopson. *Arms and Armor of the Samurai.* Crescent Books, New York, 1996.

Breen, John and M. Teeuwen. *Shinto in History: Ways of the Kami.* University of Hawaii Press, Honolulu, 2000.

Breen, John and M. Teeuwen. *A New History of Shinto.* Wiley-Blackwell Press, Oxford, 2010.

Bronowski, Jacob. *Science and Human Values.* Faber and Faber, Ltd., London, 2011.

Brown, Delmer M. and Ichiro Ishida. *The Future and the Past: A Translation and Study of the Gukansho, an Interpretive History of Japan Written in 1219 by Jien.* University of California Press, Berkeley, 1979.

Brown, Kendall H. (editor). *Japanese Warriors, Rogues and Beauties: Woodblocks From Adventure Stories.* Dover Publications, Inc., New York, 2010.

Brownlee, John S. *Political Thought in Japanese Historical Writing: From Kojiki (712) to Tokushi Yoron (1712).* Wilfred Laurier University Press, Ontario, 1991.

Brownlee, John S. *Japanese Historians and the National Myth, 1600-1945: The Age of the Gods and Emperor Jimmu.* University of British Columbia Press, Vancouver, 1997.

Bryant, Anthony. *Early Samurai 200-1500 A.D.* Osprey Publication, Oxford, 1991.

Bryant, Anthony. *Samurai 1500-1600.* Osprey Publication, Oxford, 1994.

Chiba, Reiko. *The Japanese Fortune Calendar.* Charles Tuttle Co., North Clarendon,1967.

Cooke, Tim (editor). *Concise History of World Religions: An Illustrated Time Line.* Washington, D.C., 2011.

Culture Grams Vol. 4: "Asia and Oceana." Brigham Young University Press, Provo, 2008.

Davis, F. Hadland. *Myths and Legends of Japan.* Dover Publications, New York, 1992.

Derr, Mark. *The Frontiersman: The Real Life and the Many Legends of Davy Crockett.* William Morrow and Co., New York, 1993.

DeVinne, Theo. L. *The Invention of Printing.* Francis Hart & Co., New York, 1876.

Dorson, Richard M. *Folk Legends of Japan.* Charles Tuttle Co., North Clarendon, 1962.

Dundes, Alan (editor). *The Study of Folklore.* Prentice-Hall, Inc., Englewood Cliffs, 1965.

Edmonds, Michael. *Out of the Northwoods: The Many Lives of Paul Bunyan.* Wisconsin Historical Society Press, Madison, 2009.

Embree, John F. *The Japanese Village.* Routledge, Trench, Trubner and Co., Oxford, 1946.

Flags of the World. 9 Aug., 2003. *Japan: An Illustrated Encyclopedia Vol. 2 (M-Z): national flag.* June 2004 <http://www.1uptravel.com/flag/flags/jp.html>

Fox-Davies, A. C. *A Complete Guide to Heraldry.* T.C. And E.C. Jack, London, 1909.

Friday, Karl. *The History Teacher* 27, No. 3: pp. 339-49. "Bushido or Bull? A Medieval Historian's Perspective On The Imperial Army And the Japanese Warrior Traditions". May 1994).

Goldsack, Gaby (compiler). *Flags of the World.* Parragon Publishing, Bath, 2004.

Gordon, W. J. *Flags of the World, Past and Present; Their Story and Associations.* Frederick Warne and Co., London, 1915.

Graf, John. *World War 2 Collectibles.* Krause Publications, Iola, 2007.

Greve, Gabriele. The Daruma Museum. <www.http://darumasan.blogspot.com/>

Greve, Gabriele. <www.http:darumamuseum.blogspot.com/2007/04/bishamonten-festival.html>

Guide to Flags of the World. Firefly Books, New York, 2003.

Guth, Christine. *Art of Edo Japan: The Artist and the City 1615-1868.* Yale University Press, New Haven, 1996.

Hadamitzky, Wolfgang and Mark Spahn. *Kanji and Kana.* C. E. Tuttle Publishing Co., Tokyo, 1981.

Hadamitzky, Wolfgang. *Kanji and Kana: A Handbook of the Japanese Writing System.* C. E. Tuttle Publishing Co., North Clarendon, 1997.

Hambly, Wilfred Dyson. *The History of the Tattoo.* Dover Publications, New York, 2009.

Hammel, Eric. *Air War Pacific Chronology: America's Air War Against Japan in East Asia and the Pacific 1941-1945.* Pacifica Press, Pacifica, 1988.

Hansen, Chad. *Tao Te Ching on the Art of Harmony.* Duncan Baird Publishers, Ltd., London, 2009.

Harris, Ian. (editor). *Buddhism and Politics In Twentieth-Century Asia.* London, 1999.

Havens, Norman. "Shinto." (*Nanzan Guide to Japanese Religions*). University of Hawaii Press, Honolulu, 2006.

Hay, Jeff (editor). *Shinto.* Greenhaven Press, New York, 2006.

Henshall, Kenneith G. *A History of Japan: From the Stone Age to Superpower.* St. Martin's Press, Inc., New York, 1999.

Hewitt, Mike. *Uniforms and Equipment of the Imperial Japanese Army in World War 2.* Schiffer Publishing, Atglen, 2002.

Hockley, Allen. *Chikanobu, Modernity and Nostalgia in Japanese Prints* (Coats, Bruce A.*),* "New Age Warriors: Redeploying the Heroic Ethos in the late Meiji Period." Hotei Publishing, Leiden, 2006.

Holtom, Daniel. *The Meaning of Kami in Japanese Religions: Past and Present.* University of Hawaii Press, Honolulu, 1977-1979.

Hu, Kemin. *The Suyuan Stone Catalogue: Scholars' Rocks in Ancient China.* Weatherhill Publishers, Boston, 2002.

Huish, Marcus Bourne. *Japan and Its Art.* Routledge, Oxford, 1889.

Hunter, Jack. *Samurai & Tiger Wars: Art by Kuniyoshi & Others.* Shinbaku Books, 2013.

Inazo, Nitobe. *Bushido-The Soul of Japan.* C. E. Tuttle Publishing Co., North Clarendon, 1969.

Inoue, Nobutaka and Helen Hardacre. *History of Religions* 27, No. 3: pp. 326-353. "The Shinto World Of The 1880s: Sano Tsunehiko's 'A Journey To The East'", 1988.

Inoue, Nobutaka. *Shinto: A Short History*. Routledge Publishing Co., Ltd., Oxford, 2004.

Jansen, Marius. *The Making Of Modern Japan*. Harvard University Press, Cambridge, 2000.

Jones, Landon Y. *Great Expectations: America and the Baby Boom Generation*. Ballantine Books, New York, 1986.

Jones, Marc Edmund. *Problem Solving By Horary Astrology*. Sabian Publishing Society, Evanston, 1971.

Kak, Subhash. *Brahmavidya: The Adyar Library Bulletin*, pp. 1-17, "The Vedic Gods of Japan". 2004.

Kaneda, Fujihiko. *Easy Kanji*. Passport Books, Chicago, 1996.

Kasulis, Thomas P. *Shinto: The Way Home*. University of Hawaii Press, Honolulu, 2004.

Kato, Genchi. *A Study of Shinto: The Religion of the Japanese Nation*. Barnes & Noble Books, New York, 1971.

Kemp, Peter. *Oxford Dictionary of Literary Quotations*. Oxford University Press, Oxford, 1999.

Keene, Donald. *Major Plays of Chikamatsu*. Columbia University Press, New York, 1961.

Kim, Jeongjin, Xiaomei Zhang and Paul Wyeth. *e-Preservation Science*, No. 5: pp. 41-46, "The Inherent Acidic Characteristics of Aged Silk", 2008.

Kitagawa, Joseph. *On Understanding Japanese Religion*. Princeton University Press, Princeton, 1987.

Kitchener, Andrew C. and Andrew J. Dugmore. *Animal Conservation Forum*, Vol. 3, No. 2: pp. 113-124. "Biogeographical Change in the Tiger, Panthera tigris", May 1, 2000.

Kitchener, Andrew C. and Nobuyuki Yamaguchi. *Tigers of the World: The Science, Politics and Conservation of Panthera tigris*, 2nd Edition, pp 53-84, "What is a Tiger? Biogeography, Morphology and Taxonomy." Amsterdam, 2010.

Lee, Sherman E. *A History of Far Eastern Art*. Prentiss-Hall, Inc., Upper Saddle River, 1984.

Levy, Ian Hideo. *The Ten Thousand Leaves: A Translation of the Manyoshu, Japan's Premier Anthology of Classical Poetry*. Princeton University Press, Princeton, 1987.

Lim, David (editor). *Sharing Jesus Effectively in the Buddhist World*. William Carey Library Publishers, Pasadena, 2005.

Lin, Derek. *Tao Te Ching*. Skylight Paths Publishing, Nashville, 2011.

Lin, Derek. *The Tao of Joy Every Day*. Penguin Group, New York, 2011.

Linhart, Sepp and Sabine Fruhstuck (editors). *The Culture of Japan as Seen Through its Leisure*. State University of New York Press, New York, 1998.

Littleton, C. Scott. *Shinto*. Oxford University Press, Oxford, 2002.

MacArthur, Robert H., and Edward O. Wilson. *The Theory Of Island Biogeography*. Princeton University Press, Princeton, 1967.

Mackenzie, Donald A. *Myths of China and Japan*. The Grasham Publishing Co., Ltd., London, 1923.

Mason, Penelope. *History of Japanese Art*. Pearson Prentice Hall, Upper Saddle River, 2004.

Matsumura, Kazuo. *Mythical Thinkings: What Can We Learn from Comparative Mythology?* Countershock Press, Australia, 1987-2013.

Meiri, Shai, Natalie Cooper and Andy Purvis. *Proceedings of the Royal Society of London*, Series B 275, 141-48, "The Island Rule: Made To Be Broken?", 2008.

Miner, E. R., Robert E. Morrell, and Hiroko Odogiri. *The Princeton Companion to Classical Japanese Literature*. Princeton University Press, Princeton, 1985.

Moore, Jennifer. *Confucianism: Origins, Beliefs, Practices, Holy Texts, Sacred Places*. Oxford University Press, New York, 2002.

Morimoto, Toyotomi. *Japanese Americans and Cultural Continuity: Maintaining Language and Heritage*. Garland Publishing, New York, 1997.

Nakao, Seigo. *Random House Japanese-English English-Japanese Dictionary.* Random House Publishing, New York, 1997.

Nakata, Tadao and Thomas B. Nelson. *Imperial Japanese Army and Navy Uniforms & Equipment.* Ironside International Publishers, Inc., Nanjemoy, 1975.

Neuer, Roni and Herbert Libertson. *Ukiyo-e: 250 Years of Japanese Art.* Gallery Books, New York, 1979.

Nihon Fuzokushi Gakkai (Society for the Study of Japanese Historical Customs), ed., *Nihon fuzoku shi jiten* (Encyclopedia of Japanese Historical Customs), 365. Kobundo, Tokyo, 1979.

Nosco, Peter (editor). *Confucianism and Tokugawa Culture.* University of Hawaii Press, Honolulu, 1997.

Omura, Hiroshi. *Mountain Research and Development* 24(2): 179-182, "Trees, Forests and Religion in Japan". 2004.

Ono, Sokyo. *Shinto: The Kami Way.* Charles E. Tuttle, Co., North Clarendon, 1993.

Pennington, Lee. *Seventh Asian Studies Conference* "Stitches and Sutures: Thousand-Stitch Belts, Military Medicine, and Soldiers' Bodies in Wartime Japan." Ichigaya Campus of Sophia University, Japan. June 21-22, 2003.

Pennington, Lee. *Casualties of History: Japanese Servicemen and the Second World War.* Cornell University Press, Ithaca, 2015.

Picken, Stuart D. B. *Historical Dictionary of Shinto,* 2nd edition. Scarecrow Press, Lanham, 2011.

Piggott, Juliet. *Japanese Mythology.* Hamlyn Publishing Group, New York, 1969.

Pitelka, Morgan. *Spectacular Accumulation: Material Culture, Tokugawa Ieyasu, And Samurai Sociability.* University of Hawaii Press, Honolulu, 2016.

Quebec Aikibudo Federation. *Animal Art in Japanese Sword Culture.* 29 July 2004. 29 Sept. 2004 <http:www3.sympatico.ca/apc/entsuba.htm>

Ratti, Oscar and Adele Westbrook. *Secrets of the Samurai: Martial Arts of Feudal Japan.* Diane Publishing Co., Darby, 1999.

Reader, Ian. *Religion in Contemporary Japan.* University of Hawaii Press, Honolulu, 1991.

Renard, John. *101 Questions and Answers on Confucianism, Daoism, and Shinto.* Paulist Press, New York, 2002.

Renard, John. *The Handy Religion Answer Book.* Visible Ink Press, Canton Charter Township, 2012.

Rice, Don (editor). *Animals: A Picture Sourcebook.* Van Nostrand Reinhold Co., New York, 1979.

Richie, Donald and Alexandre Georges. *The Temples of Kyoto.* Charles E. Tuttle Publishing Company, Rutland, 1995.

Roberts, Luke. *Performing The Great Peace: Political Space & Open Secrets In Tokugawa Japan.* University of Hawaii Press, Honolulu, 2012.

Sansom, George. *A History of Japan: 1615-1867.* Stanford University Press, Stanford, 1963.

Schmorleitz, Morton S. *Castles In Japan.* Charles E. Tuttle Company, Tokyo, 1974.

Sesko, Markus. *Identifying Japanese Style Script.* Lulu.com Publishing, 2014.

Shirane, Haruo. *Traces of Dreams: Landscape, Cultural Memory, and the Poetry of Basho.* Stanford University Press, Stanford, 1998.

Skulski, Janus and Stefan Draminski. *Battleships Yamato and Musashi.* Bloomsbury Publishing, London, 2017.

Smith, Sir Grafton Elliott. *The Evolution of the Dragon.* Forgotten Books Publishing, London, 2008.

Smith, Huston. *The World's Religions.* Harper Collins Publishers, New York, 1991.

Smith, Whitney. *Flags Through the Ages and Across the World.* McGraw-Hill, New York, 1975.

Smithsonian Institution. *Masterpieces of Chinese and Japanese Art: Freer Gallery of Art Handbook.* Garamond/Pridemark Press, Baltimore, 1976.

Smyers, Karen Ann. *The Fox and the Jewel: Shared and Private Meanings in Contemporary Japanese Inari Worship.* University of Hawaii Press, Honolulu, 1999.

Stanley-Baker, Joan. *The Transmission of Chinese Idealist Painting to Japan.* University of Michigan Press, Ann Arbor, 1992.

Stern, Harold P. *Birds, Beasts, Blossoms, and Bugs: The Nature of Japan.* Harry N. Abrams, Inc., New York, 1976.

Stewart, Daniel Blair. *New Asia*. Lulu.com Publishing, 2016.

Sun, Ruth Q. *The Asian Animal Zodiac*. Book Sales, Inc., New York, 1974.

Suzuki, Yui. *Medicine Master, Buddha: The Iconic Worship of Yakushi in Heian Japan*. Brill Publishers, Leiden-Boston, 2012.

Tadatoshi, Fujii. *Soldier's War*. Asahi Shimbun, Tokyo, 2000.

Takahashi, Seiichiro. *Traditional Woodblock Prints of Japan*. John Weatherhill, Inc., 1972.

Takumi, Hideo et al. *Meiji Taisho no bijutsu* (Art of the Meiji and Taisho Periods), pp. 115-117. Yubikaku, Tokyo, 1983.

Temple, Charles R. *Traditional Themes in Japanese Art*. Regent Press, Berkeley, 2008.

Thompson, Susan. *Annals of the New York Academy of Sciences*, No. 314: "Paper Manufacturing and Early Books," pp. 167-176. 1978.

Tilson, Ronald and Philip Nyhus (editors). *Tigers of the World: The Science, Politics and Conservation of Panthera tigris*. Academic Press, Cambridge, 2010.

Tsien, Tsuen-Hsuin. *Science and Civilization in China, Chemistry and Chemical Technology* (Joseph Needham), Vol. 5, Part 1 "Paper and Printing." Cambridge University Press, Cambridge, 1985.

Tujii, T. *The Japanese Trees*. Tokyo, 1995.

Turnbull, Stephen. *Samurai Warlords-The Book of Daimyo*. Villiers House, Warwickshire, 1992.

Turner, Alan. *The Big Cats and Their Fossil Relatives: An Illustrated Guide to Their Evolution and Natural History*. Columbia University Press, New York, 1997.

Tyler, Royall. *Japanese Tales*. Pantheon Press, New York, 2002.

Tzu, Lao and Muller, Charles (translator). *Tao Te Ching*. Barnes & Noble Books, New York, 2005.

Tzu, Lao and Lin, Derek (translator). *Tao Te Ching*. Skylight Paths Publishing, Nashville, 2011.

Ueda, Mokoto. *Basho and His Interpreters: Selected Hokku with Commentary*. Stanford University Press, Stanford, 1992.

Van Gulik, Robert. *The Monkey and the Tiger*. The University of Chicago Press, Chicago, 1965.

Varley, H. Paul. *Jinno Shotoki: A Chronicle of Gods and Sovereigns*. Columbia University Press, New York, 1980.

Vuori, J., and S. Tse. K. Mackay (editor). *Preliminary Study of a Microsection Method For Measuring the pH of Textiles*, pp. 63-73, Textile Specialty Group Postprints, American Institute For Conservation, Washington, D.C., 2005.

Wakagashira. 2006. Aug. 2005 <http://www.wakagashira.com/history/haramaki>

Walker, Brian Browne. *The I Ching or Book of Changes*. St. Martin's Press, New York, 1992.

Wang, Chong. *Interpreting Zheng Chengong: the politics of dramatizing a historical figure in Japan, China, and Taiwan (1700-1963)*. University of Hawaii, Honolulu, 2007.

Watanabe, Kazuhiro. *Bulletin of the National Museum of Japanese History*, 174, "Study of Samuhara Belief: Transformation From Protection Against Injuries to Protection Against Bullets," pp. 95-118. 2012.

Werner, Edward T. C. *Myths and Legends of China*. G. G. Harrap & Co., Ltd., London, 1922.

Wilkinson, Endymion. *Chinese History: A New Manual*. Harvard University Asia Center for The Harvard-Yenching Institute, 2012.

Wong, Aida-Yuen. *Artibus Asiae*, 60, 2: "A New Life For Literati Paintings in the Early Twentieth Century: Eastern Art and Modernity, A Transcultural Narrative." pp. 297-326. 2000.

Yomiuri Shimbun, The. "Kaminari mon ni yon shojo 'senninbari' no negai-chimata ni mitsu engo no mago koro" (A request for senninbari stitches by four young girls at the Kaminari Gate: a public overflowing of true feelings of support), Dec. 2, 1931: 2.

Yomiuri Shimbun, The. "Ryuko okure no sennin-musubi." (The belated fad of thousand-knot belts), July 1, 1904: 3.

Yoshimura, Akira. *Build The Musashi: The Birth and Death of the World's Greatest Battleship*. Kodansha Publishing, Tokyo, 1991.

Znamierowski, Alfred. *The World Encyclopedia of Flags*. Lorenz Books, London, 1999.

A Camera Trip Through Camp Maxey: A Picture Book of the Camp And Its Activities. The Ullman Company, Brooklyn.

www.http://omamorifromjapan.blogspot.com/2011/08/tora-tiger-toys.html

www.http://onmarkproductions.com

Bibliography: Conservation and Preservation of Textiles

Applebaum, Barbara. *Conservation Treatment Methodology*. Create Space: Independent Publishing Platform; 1st Edition, 2011.

Ashley-Smith, Jonathan. *Risk Assessment for Object Conservation*. Routledge Publishing, Oxford, 1999.

Bachmann, Konstanze. *Conservation Concerns: A Guide for Collectors and Curators*. Smithsonian Books, Washington, D. C., 1992.

Back, E. A. *Farmer's Bulletin* No. 1353, "Clothes Moths and Their Control". U. S. Department of Agriculture, 1935.

Back, E. A. and R. T. Cotton. *Refrigerating Engineering* 13: pp. 365-6, "Effect of Cold Storage upon Clothes Moths". 1927.

Bio-Integral Resource Center (BIRC). *1997 Directory of Least Toxic Pest Control Products*. IPM Practitioner 18 (11/12): pp. 1-39.

Boersma, Foekje et.al. *Unraveling Textiles: A Handbook for the Preservation of Textile Collections*. Archetype Books, Oxford, 2013.

Brooks, Mary and Dinah Eastop. *Changing Views of Textile Conservation*. Getty Conservation Institute, Los Angeles, 1st Edition, 2011.

Bry, R. E., L. L. McDonald and J.H. Lang. *Journal of Economic Entomology* No. 65(6) pp. 1735-1736. "Protecting Stored Woolens Against Fabric-Insect Damage: A Long-Term Nonchemical Method".

Elledge, Harvey G. and Alice Wakefield. *The Conservation of Textiles*. Nabu Press, Charleston, 2011.

Fields, P. G. *Journal of Stored Products Research*. No. 28: pp. 89-118 "The Control of Stored-Product Insects and Mites with Extreme Temperatures". 1992.

Finch, Karen. *Care and Preservation of Textiles*. Lacis Publishing, Berkeley, 1991.

Florian, M. – L. E. *Leather Conservation News* No. 3: pp. 1-13, "The Freezing Process-Effects on Insects and Artifact Materials". 1986.

Florian, M. – L. E. *Collection Forum* No. 6: pp. 1-7 "Freezing for Museum Insect Pest Eradication". 1990.

Haskins, Scott. *How to Save Your Stuff from Disaster*. Preservation Help Publications, Washington, D. C., 1996.

Hayward, Maria and Elizabeth Kramer. *Textiles and Text: Reestablishing the Links between Archival and Object-based Research*. Archetype Books, London, 2007.

Horie, C. V. *Materials for Conservation*. Routledge Publishing, Oxford, 2010.

Houck, M. M. *Identification of Textile Fibers*. Woodhead Publishing, Cambridge, 2009.

Janaway, Robert and Paul Wyeth. *Scientific Analysis of Ancient & Historic Textiles*. Archetype Books, Oxford, 2007.

King, Rosalie. *Textile Identification, Conservation, and Preservation*. Noyes Publications, Norwich, 1985.

Landi, Sheila. *The Textile Conservator's Manual, 2nd Edition*. Reed Education & Professional Publishing, Ltd., Oxford, 1999.

Leene, Jentina. *Textile Conservation*. Butterworth & Co. Publishers, Ltd., Oxford, 1972.

Lennard, Frances. *Textile Conservation (Conservation Museology)*. Taylor & Francis Publishers, Oxford, 2010.

Mailand, Harold F. *Preserving Textiles: A Guide for the Non-Specialist.* Indianapolis Museum of Art, 1999.

Mallis, A. *Handbook of Pest Control.* Franzak and Foster, Cleveland, 1982.

Needles, Howard L. *Textile Fibers, Dyes, Finishes and Processes: A Concise Guide.* 1st edition, William Andrew, New York, 1987.

Olkowski, W., S. Daar and H. Olkowski. *Common-Sense Pest Control: Least-Toxic Solutions for Your Home, Garden, Pets, and Community.* Taunton Press, Newtown, 1991.

Ordonez, Margaret T. *Your Vintage Keepsake: A CSA Guide to Costume Storage and Display.* Texas Tech University Press, Lubbock, 2001.

Phipps, Elena. *Looking at Textiles: A Guide to Technical Terms,* 1st edition. J. Paul Getty Museum, Los Angeles, 2012.

Richmond, Alison and Alison Bracker. *Conservation.* Routledge Publishing, Oxford, 2009.

Rust, M. K., and J. M. Kennedy. *The Feasibility of Using Modified Controlled Atmospheres to Control Insect Pests in Museums.* Getty Conservation Institute, Marina del Rey, 1992.

Smith, Joyce A., Ph.D. *Ohio State University Extension Fact Sheet.* 7 Mar. 2002. "Rayon–The Multi-Faceted Fiber". Feb. 2003. http://ohioline.osu.edu/

Staniforth, Sarah. *Historical Perspectives on Preventive Conservation, 1st edition.* Getty Conservation Institute, Los Angeles, 2013.

Strang, T. J. K. *Collection Forum* 8, pp. 41-67, "A Review of Published Temperatures for the Control of Pest Insects in Museums". 1992.

Temple, Mary Beth. *Rescuing Vintage Textiles.* Saint Johann Press, Haworth, 2000.

The Science for Conservators Series, Volume 1: An Introduction to Materials (Heritage: Care-Preservation-Management). Routledge Publishing, Oxford, 1992.

Timar-Balazsy, Agnes and Dinah Eastop. *Chemical Principles of Textile Conservation.* Routledge Publishing, Oxford, 1998.

http://www.bt.cdc.gov/disasters/mold/protect.asp

http://bugs.osu.edu/schoolipm/IPMfiles/IPM Manual/EPA IPMmanual chap-7.pdf

http://conservationresources.com/

http://conservation-us.org

http://conservation-wiki.com

http://www.epa.gov/mold/moldguide.html

http://www.extension.missouri.edu/p/GH598

http://saveyourself.com/blog/

http://www.si.edu/mci/english/research/index.html

http://www.si.edu/mci/english/learn more/taking-care/mnm.html

http://talasonline.blogspot.com/2010/01/buffered-vs-unbuffered-tissue.html

Appendix

1. p.16: Glacier cover in Japan (Hondo, and Hokkaido Island), at the height of the last glaciation about 20,000 years ago.

   Map of the Periglacial Region in Japan at the Height of the Last Glaciation about 20,000 years ago. (No changes made to map). Map credit/attribution: Davison A, Chiba S, Barton NH, Clarke B. *PloS Biology* No. 3 (9, e282). doi:10.1371/journal.pbio.0030282 [1] "Speciation and Gene Flow between Snails of Opposite Chirality" (in English), 2005.

2. p.17: Japan Country Profile

   Courtesy of www.Mapsopensource.com (No changes made to map).

3. p.35: Miharu Hariko:

   Some sources state that the Miharu Hariko initially began after the early-Edo period Miharu-han Lord, Akita Yoshisue brought home doll craftsmen from Edo following his San kin-kotai, as a venture to promote non-farming industry during the off-season.

4. p.50: *Umi Yukaba* Song Lyric:

   "If I go away to the sea,
   I shall be a washed up corpse

   If I go away to the mountain,
   I shall be a corpse in the grass.

   But if I die for the Emperor,
   It will not be a regret!"

5. p.109: The Lotus *Sutra:*

   Ch. 1, Introduction
   Ch. 2, Ways and Means
   Ch. 3, A Parable
   Ch. 4, Faith and Understanding
   Ch. 5, Parable of the plants
   Ch. 6, Assurances of Becoming a Buddha
   Ch. 7, The Magic City
   Ch. 8, Assurances for 500 *Arhats*
   Ch. 9, Assurances for the Trainees and Adepts.
   Ch. 10, Teacher of the Dharma
   Ch. 11, The Treasure *stupa*
   Ch. 12, *Devadatta*
   Ch. 13, Encouragement to uphold the sutra
   Ch. 14, Peace and Contentment

Ch. 15, Springing Up from the Earth
Ch. 16, The eternal lifespan of the *Tathagata*
Ch. 17, Merits and Virtues of enlightenment
Ch. 18, Merits and Virtues of Joyful Acceptance
Ch. 19, Merits and Virtues obtained by a Teacher of the Dharma
Ch. 20, The Bodhisattva *Sadāparibhūta*
Ch. 21, The Spiritual Power of the *Tathagata*
Ch. 22, The Passing of the Commission
Ch. 23, The Bodhisattva *Bhaisajyarāja*
Ch. 24, The Bodhisattva *Gadgadasvara*
Ch. 25, The Bodhisattva *Avalokiteśvara*
Ch. 26, *Dhāranī*
Ch. 27, King Wonderfully Adorned
Ch. 28, Encouragement of the Bodhisattva *Samantabhadra*

6. p.127: Cintamani or Chintamani Jewel:

These wish-fulfilling jewels/stones are connected to the drops of pure water painted on the side of the water vase next to the mantra, Samuhara. Depending upon the wishes of the believer, the stones may be used to purify water and remove sickness and suffering. They represent the teachings of the Buddha and reflect his highest virtues.

7. p.142: Memorial Day: *Bedroom Eyes*

A short story written about one of the crew members of the B-29 Superfortress *Bedroom Eyes* by his granddaughter, Lori Janeski. It appeared in the publication *The Catholic Geeks* on May 20, 2015.
https://libertyislandmag.com/2015/05/20/memorial-day/

8. p.151: The *Sutra* to the Buddha of Healing –Yakushi Nyorai:

1: May a radiant light blaze forth from my body after enlightenment, brightening countless realms, and may all beings have perfect physical form, identical to my own. 2: May my body be like pure radiant lapis lazuli, with a radiance more brilliant than the sun and moon, illuminating all who travel in darkness, enabling them to tread upon their paths. 3: By my limitless insight and means, may I enable all beings to obtain the necessities of life. 4: May all beings be shown the path of enlightenment and may adherents to the *sravaka* or *pratyekabuddha* paths become established in *Mahayana* practices. 5: May all beings be aided to follow the precepts of moral conduct. After hearing my name, those who have broken the precepts will be aided to regain their purity and prevented from sinking to a woe some path of existence. 6: May all who are deformed or handicapped in any way have their deformities removed upon hearing my name. 7: May all who are ill be cured upon hearing my name. 8: May all sentient beings who are restrained by their circumstances of birth find a favorable rebirth and progress towards Liberation. 9: May all who are caught in Mara's net, entangled in negative views, be caused to gain correct views and thus practice the Bodhisattva Way. 10: May all who are punished by the king be freed of their troubles. 11: May those who are desperately famished be given food. May they ultimately taste the sublime Teachings. 12: May all who are destitute of clothes obtain attractive garments and various adornments upon concentrating on my name.

9. p.153: The Names Of The Twelve Divine Generals-Protectors of the Yakushi Nyorai:

1: Bikara Taisho
2: Shotora Taisho
3: Shindara Taisho
4: Makora Taisho
5: Haira Taisho
6: Indara Taisho
7: Sanchira Taisho
8: Anira Taisho – Associated with the Dragon
9: Anchira Taisho
10: Mekira Taisho – Associated with the Tiger
11: Basara Taisho
12: Kubira Taisho

Acknowledgments

I would like to thank the individuals listed here who helped me to make this book possible. I would like to extend a special thanks to Takehito Jimbo whose willingness to help me with the translation work, at any time of the day or night, was invaluable. There are undoubtedly those whose names I may have forgotten to include, and for that I apologize.

Greg Babich
Dolan Bortner
John* and Rumiko* Byrum
Eric Doody
Travis Duffy
John Egger*
Ken Emery
Carla L. Funk
Ruth Funk Center for the Textile Arts, Melbourne, Florida
Matthew Hecker
Takehito Jimbo
Dean Kanabrocki
Thomas Prall
Tomas Termote
Stephen Wallace, D.D.S.
Vintage Productions
Ron Werneth

*(deceased)

About the Author

Michael Bortner has been a collector of Imperial Japanese *hinomaru yosegaki* for over fifty years. His authoritative volume on that subject, as well as on *senninbari,* was released in 2008. This book is intended to provide the reader with an equally thorough and detailed discussion on the history and use of the tiger by the Japanese to decorate a rare few of their good luck flags. Drawing upon his own extensive collection, as well as those from fellow collectors worldwide, the author has assembled the first book in English solely on the subject.

The author earned his B. A. Degree in History from the University of California, Irvine, and his M. A. Degree in Anthropology from the California State University, Hayward. He earned his Doctor of Dental Surgery from Georgetown University, Washington, D. C., and currently lives in the San Francisco Bay area with his wife Andrea; they have two children.

Michael Bortner may be contacted by writing to:
P. O. Box 927, Belvedere-Tiburon, CA 94920.

www.ingramcontent.com/pod-product-compliance
Ingram Content Group UK Ltd.
Pitfield, Milton Keynes, MK11 3LW, UK
UKRC031012141026
15625UKWH00082B/658

* 9 7 8 1 9 4 3 4 9 2 5 7 2 *